Climate Change and Tradition in a Small Island State

T0331271

The citizens of the Marshall Islands have been told that climate change will doom their country, and they have seen confirmatory omens in the land, air, and sea. This book investigates how grassroots Marshallese society has interpreted and responded to this threat as intimated by local observation, science communication, and Biblical exegesis. With grounds to dismiss or ignore the threat, Marshall Islanders have instead embraced it; with reasons to forswear guilt and responsibility, they have instead adopted in-group blame; and having been instructed that resettlement is necessary, they have vowed instead to retain the homeland. These dominant local responses can be understood as arising from a pre-existing, vigorous constellation of Marshallese ideas termed "modernity the trickster": a historically inspired narrative of self-inflicted cultural decline and seduction by Euro-American modernity. This study illuminates islander agency at the intersection of the local and the global, and suggests a theory of risk perception based on ideological commitment to narratives of historical progress and decline.

Peter Rudiak-Gould is a Mellon Postdoctoral Fellow in the Department of Anthropology at McGill University.

Routledge Studies in Anthropology

1. **Student Mobility and Narrative in Europe**
   The New Strangers
   Elizabeth Murphy-Lejeune

2. **The Question of the Gift**
   Essays across Disciplines
   Edited by Mark Osteen

3. **Decolonising Indigenous Rights**
   Edited by Adolfo de Oliveira

4. **Traveling Spirits**
   Migrants, Markets and Mobilities
   Edited by Gertrud Hüwelmeier and Kristine Krause

5. **Anthropologists, Indigenous Scholars and the Research Endeavour**
   Seeking Bridges Towards Mutual Respect
   Edited by Joy Hendry and Laara Fitznor

6. **Confronting Capital**
   Critique and Engagement in Anthropology
   Edited by Pauline Gardiner Barber, Belinda Leach and Winnie Lem

7. **Adolescent Identity**
   Evolutionary, Cultural and Developmental Perspectives
   Edited by Bonnie L. Hewlett

8. **The Social Life of Climate Change Models**
   Anticipating Nature
   Edited by Kirsten Hastrup and Martin Skrydstrup

9. **Islam, Development, and Urban Women's Reproductive Practices**
   Cortney Hughes Rinker

10. **Senses and Citizenships**
    Embodying Political Life
    Edited by Susanna Trnka, Christine Dureau and Julie Park

11. **Environmental Anthropology**
    Future Directions
    Edited by Helen Kopnina and Eleanor Shoreman-Ouimet

12. **Times of Security**
    Ethnographies of Fear, Protest and the Future
    Edited by Martin Holbraad and Morten Axel Pedersen

13. **Climate Change and Tradition in a Small Island State**
    The Rising Tide
    Peter Rudiak-Gould

Climate Change and Tradition in a Small Island State

The Rising Tide

Peter Rudiak-Gould

Routledge
Taylor & Francis Group
LONDON AND NEW YORK

First published 2013
by Routledge
711 Third Avenue, New York, NY 10017

Simultaneously published in the UK
by Routledge
2 Park Square, Milton Park, Abingdon, Oxfordshire OX14 4RN

Routledge is an imprint of the Taylor and Francis Group, an informa business

First issued in paperback 2015

Library of Congress Cataloging-in-Publication Data

Rudiak-Gould, Peter.
Climate change and tradition in a small island state : the rising tide / by Peter Rudiak-Gould.
pages cm. — (Routledge studies in anthropology ; 13)
Includes bibliographical references and index.
1. Climatic changes—Marshall Islands. 2. Climatic changes—Social aspects—Marshall Islands. 3. Marshall Islands—Environmental conditions. 4. Marshall Islands—Social life and customs. 5. Environmental degradation—Marshall Islands. 6. Blame—Social aspects—Marshall Islands. 7. Risk perception—Marshall Islands. 8. Social change—Marshall Islands. 9. Social movements—Marshall Islands. I. Title.
QC903.2.M37R83 2013
363.738'74099683—dc23
2013006447

ISBN 978-0-415-83249-6 (hbk)
ISBN 978-1-138-95281-2 (pbk)
ISBN 978-0-203-42742-2 (ebk)

Typeset in Sabon
by IBT Global.

Ij iọkwe ḷọk aelōñ ko ijo iaar ḷotak ie
Meḷan ko ie, im iaḷ ko ie, im iaieo ko ie
Ijāmin ilọk jān e bwe ijo jikū eṃool
Im aō ḷāṃoran indeeo
Eṃṃanḷọk ñe inaaj mej ie

I love the islands where I was born
The surroundings, the paths, the gatherings
I will never leave because it is my true home
And my inheritance forever
It is best that I die there

—First national anthem of the Republic of the Marshall Islands

We're now looking at a more than one meter sea level rise by the end of the century. . . . Anyone looking objectively at this region has to see the need for relocation.

—Prof. Patrick Nunn[1]

1. Journal, 2009c

In memory of Alfred DeBrum
May the sea never take your grave

Contents

Figures

Tables

Abbreviations

ADB	Asian Development Bank
CMI	College of the Marshall Islands
EPA	Environmental Protection Authority (Republic of the Marshall Islands)
GHG	Greenhouse gas
IPCC	Intergovernmental Panel on Climate Change
MEC	Marshalls Energy Company
MICS	Marshall Islands Conservation Society
MIMRA	Marshall Islands Marine Resources Authority
NGO	Non-governmental organization
NOAA	National Oceanic and Atmospheric Administration
OEPPC	Office of Environmental Planning and Policy Coordination (Republic of the Marshall Islands)
RMI	Republic of the Marshall Islands
UK	United Kingdom
UN	United Nations
UNESCO	United Nations Educational, Scientific and Cultural Organization
US	United States
USDA	United States Department of Agriculture
USP	University of the South Pacific
WUTMI	Women United Together Marshall Islands

Acknowledgments

A heartfelt *koṃṃooltata* goes to every Marshall Islander I met for welcoming me to your islands and sharing your thoughts. I would like especially to thank the following *riṃajel* for their unfailing hospitality, research assistance, information sharing, and bright ideas: Alfred and Tior DeBrum, Lisson and Elina Langidrik, Caios and Wisse Lucky, Raymond and Tenita DeBrum, Holden and Tarise Milne, Mark Stege, Biram Stege, Mike Kabua, Alson Kelen, Alfred Capelle, Tony DeBrum, Josepha Maddison, Jabukja Aikne, Steven Patrick, Ben Graham, Albon Ishoda, Milner Okney, Carlton Abon, Clarence Luther, Milton Zackios, Christian Lehman, Hilda Heine, Mona Levy-Strauss, Marie Maddison, Daisy Momotaro, Evelyn Joseph, Lisa Jeraan, and Kathy Jetnil-Kijiner. All of the women of WUTMI (www.wutmirmi.org) showed true kindness in allowing me to attend and participate in their climate change outreach activities. Staff of the *Marshall Islands Journal* (www.marshallislandsjournal.com) generously allowed me to reprint political cartoons.

Others, not Marshallese but with an abiding interest in the islands, were enormously helpful to me: Byron Bender, Jack Niedenthal, Scott Stege, Giff Johnson, Tamara Greenstone, Lauren Pallotta, Ingrid Ahlgren, Steve Why, Eric and Barbara Fisher, Natalie Nimmer, Erin Jacobs, Joe Genz, Todd Foster, Amy Carlson, Angela Saunders, Olai Uludong, Marilyn Low, and Kapono Ciotti. Fellow anthropologists Jessica Schwartz and Elise Berman shared some of their data with me. Julian Sachs invited me to join his scientific expedition. Conor Myhrvold graciously allowed me to include one of his photographs in this book.

I am grateful for the dedicated academic mentorship of David Gellner, Harvey Whitehouse, and Colin Scott. I am indebted to Mark Stege, Cristine Legare, Justin Barrett, and Andreas Roepstorff for reading drafts and providing feedback at many points in the process of research.

My 2007 fieldwork was supported by the Barbinder-Watson Trust Fund at St. Hugh's College, Oxford. My 2009 fieldwork was funded by the Peter Lienhardt Memorial Fund from the Institute of Social and Cultural Anthropology, Oxford, the E.O. James Bequest from All Souls College, Oxford, and by a research allowance from Jesus College, Oxford. My doctoral work was funded by the Alun Hughes Graduate Scholarship at Jesus College, Oxford, and by the Clarendon Scholarship and Overseas Research Scholarship. Further research and my 2012 field trip were made possible by the Andrew W. Mellon Foundation, with special thanks to Sarah Kaderabek.

Introduction
A Scientific Prophecy

Ujae Island is both old and new. To the anthropologist and the indigenous inhabitant, it is an ancient and permanent place, cultivated and storied by scores of generations. To the scientist it is recent, tentative—no more than a flooded reef until well into historical times, and now vulnerable to human tampering and natural variability alike. With oceans rising, this 200-acre islet, barely breaching the sea, may sink again: here and gone in a geological eyeblink—or a cultural eon.

In 2004 the specter of climate change had been far from the minds of Ujae's people, as it had been for most of the citizens of the Republic of the Marshall Islands. But by 2007 the whole world was talking about it, and some Ujae residents had begun to follow suit. Local confirmation seemed to be appearing in the form of erosion scarps and falling coconut trees. Children played on those horizontal trunks, giggling and roughhousing on evidence of their own eventual exodus. A respected local man, entering his middle years, bore witness to the changes, speaking about a particular coconut tree on a neighboring land tract that had conspicuously fallen in the last few years:

> It fell because the sea is higher. The waves knock the trees over—not just coconut trees but others as well . . . If the tide rises, it will wash all of the land and all of the trees into the ocean, and we won't be able to plant anything. God told Noah that He would never flood the earth again. But the scientists say the ocean will rise. It's hotter so the snow melts and the ocean will rise and flood the Marshall Islands. I heard this on the radio news in 2000. The ocean will rise higher than the tops of the coconut trees.

Just a few feet away, the man's 14-year-old nephew had been innocently eavesdropping on the adults, never imagining that his curiosity might deliver such news. His face turned grave. This was the boy's first exposure to the idea of climate change, the almost unimaginable notion that his entire country might become uninhabitable within his lifetime.

CLIMATE CHANGE AND THE MARSHALL ISLANDS

"The future of the Marshall Islands is in the ocean"—thus was the unintentional double entendre of the Taiwanese ambassador to the Marshall Islands as he pointed in 2005 to the country's potential for marine resource development (Journal, 2005a). This archipelago in eastern Micronesia, 3,700 kilometers southwest of Hawaii, is one of four sovereign states—the others being the Pacific countries of Kiribati and Tuvalu and the Indian Ocean atoll nation of the Maldives—facing nationwide uninhabitability, if not nonexistence, at the hands of climate change (Barnett & Adger, 2003). The best-known threat posed by climate change to this country is sea level rise. Projections for average global sea level rise by 2100 vary widely from 0.18 to 2 meters (see IPCC, 2007a, Sec. 10.6.5; Nicholls & Cazenave, 2010; Nicholls et al., 2011; Rahmstorf, 2007), but even the lower estimates are expected to threaten the habitality of low-lying islands (Barnett & Adger, 2003). Recent projections specific to the Marshall Islands predict 0.18 to 0.62 meters of sea level rise by 2090 and possibly considerably more (Australian Government, 2011: 7). The country is composed entirely of low-lying coral atolls with an average elevation of 1 meter and no point higher than 10 meters (Figures 0.1 and 0.2). On most islands it is impossible to put more than a few hundred meters between oneself and the ocean, and no place in the country is more than a kilometer inland: the 1,225 islets total a mere 181 square kilometers of land spread out over 2 million square kilometers of ocean. One of the few Marshallese communities surveyed in detail, Jeh Island in Ailinglaplap Atoll, would see up to 90% of its existing houses regularly flooded if the sea level rises by a mere 0.5 meter (Kench et al., 2011: 20). While the country's coral reefs have existed for tens of millions of years (Guilcher, 1988: 228), the islands themselves formed a mere 2,250 years ago, only a few hundred years before the first human settlers arrived (Yamaguchi et al., 2005: 31–32). The idea of an inundated Marshall Islands is therefore far from fanciful.

But sea level rise is only the most obvious and iconic climate change impact in this country. A "sinister synergy" (Patel, 2006) of processes is at work. The islands will not simply sink (RMI Government, 2006a: 28) but become more and more vulnerable to the sea. Typhoons are expected to batter the atolls with increased intensity, both in terms of rainfall and maximum wind speed (Australian Government, 2011: 6), the damage exacerbated by dying coral reefs and rising sea levels. Floods will damage infrastructure, ruin crops, and cultivate disease (Barnett & Adger, 2003: 324–326). Fresh water is naturally scarce on the islands, since there are no streams and few lakes; rainwater is unreliable during the dry season, from December to March; and groundwater is found only in thin Ghyben-Herzberg water lenses, which are easily contaminated by saltwater. Historically rainy atolls have been predicted to become drier, and vice versa (Sachs, Stege, & Keju, 2009), threatening settlement patterns based largely on historical levels of rainfall (Williamson & Sabath, 1982). Increased climatic variability will lead to more erratic tuna catches, impacting this major source of government revenue (Moss, 2007). Warming sea temperatures and ocean acidification will damage coral reefs (Australian Government, 2011: 7), jeopardizing the source of fish protein for many rural islanders and weakening one of the islands' natural buffers against sea level rise.

Human decisions and historical legacies intensify the country's vulnerability. The population has grown explosively since World War II, quintupling in fifty years (Juumemmej, 2006: 73) and now surpassing 60,000 (United Nations, 2010), many times above the traditional carrying capacity of the islands. Two thirds of the population now live in the crowded urban centers of Majuro and Ebeye (Juumemmej, 2006: 73), a result of population growth, cultural change, and displacement by American nuclear testing. Whereas Marshallese once settled on the safer tracts of land—the lagoon shore rather than the ocean, wide islands rather than narrow, the leeward end of the atoll rather than the windward—these practices have disappeared in the cities. Urban settlements now extend from the lagoon side to the ocean side, spilling to the very edge of the water (Figure 0.3). The location of Majuro, the capital city, may be the most precarious of all: the D-U-D (downtown) area sprang up on a series of long, thin islets on the exposed windward side of Majuro Atoll, in part a colonial legacy of American military involvement in the nation (Spennemann, 1996). In both Majuro and the country's other urban center, Ebeye, construction has eliminated much of the vegetation that naturally protects against wind and waves, exacerbating erosion (Juumemmej, 2006: 89, 97; Xue, 2001). In a chilling intersection of the country's nuclear past with its climatic future, rising sea levels could erode the concrete dome at Runit Island, Eniwetok Atoll, where the American military capped 84,000 cubic meters of radioactive waste left over from atomic testing (RMI Government, 2009: 13). The Marshallese government's modest means weaken hopes of technical solutions to climate change; for instance, it might be feasible to fortify the capital city against a meter of sea level rise, but such a project could cost four times the country's entire gross domestic product (Barnett, 2005). High population densities in the urban centers also stress water resources; even now most Majuro residents can access piped water only two days a week.

A vulnerable system has been defined as being "very sensitive to modest changes in climate, where the sensitivity includes the potential for substantial harmful effects, and for which the ability to adapt is severely constrained." (Crate, 2008: 571). This definition fits the Marshall Islands, as the effects of climate change are expected to be severe and the scope for adaptation has long been feared to be limited (Holthus et al., 1992). There is a very real possibility that the country will become uninhabitable in the present century (Barnett & Adger, 2003: 325–326) or during the lifetimes of Marshall Islanders living today. In this case, an entire nation would be forced to leave for other shores. It is true that thousands of Marshall

Islanders have already chosen to move to the United States, and such migration may be counted as a positive adaptation strategy (Barnett & Webber, 2010; Farbotko & Lazrus, 2012; Orlove, 2005). But mass exodus would represent an entirely different movement: a forced displacement, and an irreversible one. The legal, political, and sociocultural implications are troubling, uncertain, and in some cases unprecedented (see Adger, Barnett, & Ellemor, 2010; Campbell, 2010; Campbell, Goldsmith, & Koshy, 2005; Kempf, 2009; Larson, 2000; McAdam, 2010; Moore & Smith, 1995; Oliver-Smith, 2009; Penz, 2010; Zetter, 2010). As one journalist observed, "Unlike other refugees displaced by wars or famines, these people on the edge of the ocean face the prospect of never again having homelands to return to" (Whitty, 2003).

In this worst-case scenario, there would be no one moment of uninhabitability, no bright line at which life becomes impossible; even impossibility is a fuzzy concept. A poem by a Marshallese student refers to climate change as a "long, slow terror" (Kabua, 2010), and this is apt. Marshallese life will simply become more and more difficult, dangerous, and unpleasant, until, individual by individual, family by family, the nation has been depopulated.

An honest assessment of the climate change prognosis for a low-lying nation always involves a tension between, on the one hand, acknowledging and exhibiting the gravity of the threat and, on the other, admitting uncertainties, leaving room for hope and local agency, and avoiding the sort of speculative sensationalism that writes off atoll citizens in the name of helping them (see Barnett & Campbell, 2010; Farbotko, 2005, 2010; Farbotko & Lazrus, 2012; Mortreux & Barnett, 2009). Indeed hope may yet remain for the Marshall Islands. The natural response of a coral atoll to sea level rise and other climate change impacts is complex and poorly understood (Ford, 2012; IPCC, 2007b, Sec. 16.4.2), involving resilience as well as vulnerability (Webb & Kench, 2010). Vulnerability is not uniform across the archipelago: the country's handful of single islands—individual coral islets not connected to an atoll—enjoy slightly higher elevations and therefore far greater protection against midrange sea level rise scenarios (Kench et al., 2011: 28). A global reduction in emissions combined with locally appropriate adaptation measures may yet save low-lying island nations (Barnett & Campbell, 2010; Mortreux & Barnett, 2009: 106). Even in high-emissions scenarios, not all of the projections are unfavorable: recent predictions specific to the Marshall Islands suggest that droughts may become less common and typhoons less frequent even if more intense (Australian Government, 2011: 6). Whatever hope may remain, however, the scientific prophecy of climate change that Marshall Islanders must now live with is undeniably a dire one.

This book will describe how Marshall Islanders have interpreted and responded to this threat. My interest is the attitudes and actions of citizens, NGOs, and civil society rather than the Marshallese government. We will see the Marshallese versions of three global debates[1] about human-made climate change, namely: (1) Is it real and worthy of concern? (2) Who should bear responsibility or blame? (3) What is to be done? I have not tried to neatly separate these three interrelated debates in this book, but, roughly, Chapters 2 and 3 treat debate 1, Chapter 4 treats debate 2, and

Figure 0.1 Piñļap Island, Jaluit Atoll (Photo by the author).

Figure 0.2 Outer island shoreline dwellings, Ailinglaplap Atoll (Credit: Conor L. Myhrvold, Ailinglaplap Atoll, Marshall Islands, July 2009).

Figure 0.3 Urban center shoreline dwellings: Ocean side of the Jenrōk neighborhood, Majuro Atoll (Photo by the author).

Chapters 4 and 5 treat debate 3. Chapter 1 and 2 set the stage for this discussion: Chapter 1 describes the cultural, historical, and ideological milieu into which the idea of climate change has been received in this country, and Chapter 2 describes how Marshall Islanders become aware of the issue.

I will show that, for each debate, while agreement is not total, one particular position is noticeably dominant in Marshallese grassroots society: (1) Climate change *is* real and worthy of concern. (2) All human beings are to blame, with special attention on Marshall Islanders themselves (becoming a kind of in-group blame). (3) Regarding the *causes* of climate change, Marshall Islanders ought to pursue local mitigation[2] (reduction of their own greenhouse gas emissions) rather than protesting other countries' contributions; regarding the *impacts* of climate change, Marshall Islanders ought to adapt in situ rather than relocating en masse to another country.[3] I call these "dominant" attitudes because they are the ones that are expressed publicly, constituting the "safe" positions that will not get one into any trouble and are driving responses and outward behavior. Other answers also exist but are less commonly voiced in interviews, rarely or never expressed publicly, and lead to little or nothing in the way of responses and action. (In the next section I further discuss and defend this focus on "dominant" attitudes.)

I hope to explain why these responses are the popular ones by understanding the preexisting categories, values, and narratives that Marshall Islanders invoke to interpret climate change. Chapter 1 presents this cultural and metacultural background, in particular a constellation of local ideas that I variously call cultural decline, seductive modernity, and "modernity the trickster": an impressively mighty yet morally problematic America and its modernizing influence, powerful but untrustworthy scientists, and an ancestral Marshallese tradition under siege by internal imitation of these seductive and destructive foreign ways. These symbols and stories, tremendously influential in the Marshall Islands, go very far in elucidating why locals respond to climate change in the unique way that they do.

THEORETICAL FOUNDATIONS

In the western world, in policymaking circles, in the halls of power, climate change is usually assumed to be an "environmental issue" or a "geophysical phenomenon" (Macnaghten & Urry, 1995: 210–211; H. A. Smith, 2007: 200–201) under the academic jurisdiction of the physical sciences (Crate & Nuttall, 2009a: 394; Hassan, 2009: 42; Yearley, 2009: 390). It is the fundamental theoretical premise of this book that climate change is intimately sociocultural as well (Batterbury, 2008; Crate, 2008; Daniels & Endfield, 2008; Jasanoff, 2010; Lahsen, 2007a; Leduc, 2011; Lindisfarne, 2010; Magistro & Roncoli, 2001; Nelson, 2007; Rayner & Malone, 1998; Yearley, 2009). While anthropologists and other humanistic social scientists have largely ignored climate change for decades (Batterbury, 2008; Finan, 2007; Rayner, 1989; Townsend, 2004), recent years have witnessed a dramatic upsurge in interest (for a review see Crate, 2011). Both argument and example now show that humanistic and social scientific inquiry has an important role to play in the study of global warming. The social dimensions of climate change are myriad. The *causes* are human, stemming ultimately from a particular social and economic order (Bohren, 2009; Wilk, 2009). The *impacts* influence human actors and are observed, negotiated, and interpreted by them (Crate, 2008; Fox, 2002; Hitchcock, 2009; Jacka, 2009; Jolly et al., 2002; Lipset, 2011; Marino & Schweitzer, 2009; Thorpe et al., 2002). *Responses*, either to the science or to actual impacts, hinge on human institutions and ideologies, and themselves may have dramatic sociocultural impacts (Berkes & Jolly, 2001; Degawan, 2008; Fox, 2002; Hulme, 2009; Pettenger, 2007; Puri, 2007). The *communication* of climate change from various scientific communities to the public and from one society to another is a cross-cultural encounter and an act of cultural translation (Bravo, 2009; Hassol, 2008; Lahsen, 2010; Moser, 2010; Nerlich et al., 2010).

A second premise is the importance of the *reception*, not only the *observation*, of climate change by frontline and indigenous communities. By reception I mean the acquisition of climate change information via media, educational, governmental, NGO, or other channels that disseminate the scientific notion of anthropogenic global warming. By observation I mean the acquisition of climate change information via observation of local climate change impacts (or any other changes interpretable as such). I have elsewhere argued (Rudiak-Gould, 2011) that climate anthropology's implicit (and occasionally explicit—see Marino & Schweitzer, 2009) privileging of observation studies is problematic in many field sites, and that anthropologists would do well to adopt the approach of certain less conventional ethnographers (Lahsen, 2004, 2007b; Lazrus, 2009; Strauss, 2009; Taddei, 2009a), as well as other social scientists (Bravo, 2009; Connell, 2003; Grothmann & Patt, 2005; Hulme, 2009; Kuruppu & Liverman, 2011; Mortreux & Barnett, 2009; Paton & Fairbairn-Dunlop, 2010; Suarez & Patt, 2004; Swim et al., 2009) who take seriously the role of reception in driving climate change attitudes in "frontline" communities. I draw inspiration from ethnographic works that explore the diverse and sometimes conflicting sources of information that publics draw upon to anticipate and interpret meteorological phenomena. Depending on the society, these sources may be firsthand observation, scientific communication, and historical memory (see Lazrus [2009: 240] on Tuvalu); firsthand observation, scientific communication, and religious prophecy (see Mortreux and Barnett [2009] on Tuvalu and Gold [1998] on Rajasthan); firsthand observation, the divinations of local seers, and prophecies from both the Bible and the Koran (see Roncoli, Ingram, & Kirshen [2002] and Roncoli [2006] on African farmers); or governmental predictions and those of local weather prophets (see Taddei [2009a, 2009b] on northeast Brazil). I use the term *triangulation* (borrowing from Lazrus [2009: 24]) to describe this convergence of diverse information sources.

With these premises as foundational, I hope to unify two literatures that until now have never enjoyed the rapprochement that they deserve: the cultural theory of risk, associated in particular with Mary Douglas, and "trajectory narratives," associated with a variety of ethnographers and especially Oceanists.

Mary Douglas's cultural theory of risk (Douglas, 1982a, 1992; Douglas & Wildavsky, 1982) begins from the premise that it is impossible to possess complete knowledge of the dangers that a society faces. "Knowledge always lacks. Ambiguity always lurks" (Douglas, 1992: 9); therefore evaluating the likelihood of a particular danger can never be a perfectly value-free exercise. Even in a modernist utopia of empirical omniscience, the ranking and weighting of negative outcomes—indeed, their very characterization as *negative*—is fundamentally a question of ethics and thus of unending debate (Douglas, 1992: 31; Douglas & Wildavsky, 1982: 2). Douglas quotes philosopher Jerome R. Ravetz, who writes: "The hope that one can produce a taxonomy, evaluation, and finally a technical fix to the problems of risks is in substance as ambitious as the program of putting all of human experience and value onto a scale of measurement for mathematical or political manipulation." (Quoted in Douglas & Wildavsky, 1982: 2).

With this rationalist dream denied, risk perception is always caught up in ideology. Societies, and factions within them, attend to some hazards while ignoring others, according to whether the hazards in question "reinforce the [society's or faction's] preferred political scheme." (Douglas, 1992: 32) This selection of risks, in turn, influences how societies respond to the threat. If the threat is not selected for attention, little or no response will result. If the threat is selected for attention, the society or faction will respond in a way ideologically consistent with its social structure and associated worldview. Blame will be meted out according to the usual habits of the society: preconceived villains (depending on the society, these may be out-groups, rivals within the in-group, or the victim himself) will be deemed culpable, and appropriate action will be taken: vengeance against out-groups, calls for compensation from or punishment of an in-group rival, or acts of purification and atonement (Douglas, 1992: 5–6). As Douglas writes, "In risk perception, humans act less as individuals and more as social beings who have internalized social pressures and delegated their decision-making processes to institutions" (Douglas & Wildavsky, 1982: 80). To restate Douglas's position, what people seek to avoid is not risks per se but discourses of risk that challenge their concepts, values, political claims, and social affiliations. Risks, even those that threaten physical disaster, may indeed be *attractive* to individuals and societies as long as they flatter key ideological commitments.

Many of Douglas's more specific claims about risk perception have been roundly and convincingly critiqued. Boholm (1996) upbraids her implicit functionalism; her assumption that cosmology and sociology (or discourse and behavior) mirror each other; her old-fashioned vision of groups as discrete, homogeneous, bounded units; and the failure of her typology of societies (based on "group" and "grid" variables—see Douglas, 1970, 1982b, 1992; Douglas & Wildavsky, 1982; Thompson, Ellis, & Wildavsky, 1990) to account for more than a small percentage of the variance in individual's risk perceptions (here also see Brenot et al., 1998; though see Kahan et al., 2007 for a contrary finding). These critiques do not impede the present study. I do not assume functionalism; I assume no easy fit between cosmology and sociology (indeed, as I will explain, it is the *contrast* between discourse and behavior that is the topic and the moral heft of the discourse on which I focus); I treat "grassroots Marshall Islanders" as a bounded group not as an a priori assumption but as an a posteriori conclusion based on the near-ubiquity of a particular discourse within this population. I circumvent the critiques by discarding Douglas's *special* cultural theory of risk (group-grid theory, with its particular typology of societies and its assumptions of functionalism and boundedness) in favor of her *general* cultural theory of risk, which contends only that risk is a politicized discourse of danger, irreducible to scientific objectivity, which is not limited to being aversive for its physical impacts but also may be aversive *or attractive* for its ideological impacts.

This general cultural theory of risk is on much firmer ground. It accords well with studies in other disciplines, in particular cognitive psychology, communication, and science and technology studies—studies that have never fallen prey to the criticisms listed above. Research in several disciplines demonstrates that "pre-existing . . . norms" (Cass, 2007: 46) and ideological "prior commitments" (Jasanoff, 2010: 240) have an enormous influence on people's belief in anthropogenic global warming, their level of concern about it, assessment of the threat's severity, assignment of blame and responsibility, and preferred responses (see, for instance, Feinberg & Willer, 2011; Feygina et al., 2010; Hulme, 2010a; McCright, 2011). Such a framework fits well with literature in science and technology studies on the public uptake of scientific discourses of risk (Jasanoff & Wynne, 1998; Sarewitz, 2004; Wynne, 1992), and with literature in cognitive psychology on confirmation bias (Lord et al., 1979; Nickerson, 1998): the habit of making new ideas conform to old. It also agrees with constructivist and "epidemiology of representations" approaches to communication and cultural transmission (Sperber, 2006[1985]), which assume that a message (such as a media statement on climate change) is not a static package of information to be transferred from one party to another (as in earlier transmission [Weaver & Shannon, 1963] and memetic [Dawkins, 2006{1976}] models) but is rather a "prompt to construct meaning" (Höche, 2009: 276), reassembled by the receiver according to preexisting beliefs (Bloch, 2000; Sperber, 2000). This framework has been profitably applied to understanding the translation of climate change (Nerlich et al., 2010; Whitmarsh, 2009). The molding of climate change to fit existing concepts could be described uncharitably as intellectual laziness or prejudice-reinforcement, but perhaps it is more fruitful to see it as a creative transformation of the "impersonal, apolitical, and universal imaginary of climate change, projected and endorsed by science" into "subjective, situated and normative imaginations" (Jasanoff, 2010: 235)—the "moral reading" (Bravo, 2009: 277) of climate change so as to render it intelligible and actionable.

Which prior commitments are important for Marshallese climate change attitudes? I argue that the crucial prior commitment is the discourse of cultural decline (or "seductive modernity" or "modernity the trickster"), which I understand to be a "trajectory narrative." Trajectory narratives are discourses of the moral direction of society or the cosmos, with an associated sense of responsibility or blame for that trend. Ethnographers have shown repeatedly that narratives of this sort are key ideational resources in many societies. The literature on this topic is especially rich in the Pacific region (see in particular Ballard, 2000; Errington & Gewertz, 1986; Jorgensen, 1981; Keesing, 2000; Rudiak-Gould, 2010; Tomlinson, 2004, 2009; White, 1991), but it is by no means confined to that region (see Condry, 1976; Gold, 1998; Jackson, 2006; W. D. Smith, 2007).

In the colonial and postcolonial Pacific, the focus of these trajectory narratives is very often "tradition" (or *kastom*, or various other local equivalents): either the lamentable regression from a harmonious traditional past to a troubled modernist present or the laudable progression from a primitive traditional past to an advanced modernist present. I do not have space for a full treatment of the complex topic of "tradition" and its vast literature, but suffice it to say that I agree with and am indebted to the broad consensus of the anthropological literature. A society's notion of its "tradition" is neither a simple reflection of objective historical reality (a realist theory) nor an ungrounded, unconstrained fantasy (a radically constructivist theory) but rather the result of the creative friction between past lifeways and newer influences during the intensive cultural contact of colonialism and its aftermath (a coproduction theory). Such notions take hold of people's minds for instrumentalist reasons of political action, as well as intellectualist motives to compare past and present, foreign and native, desirable and undesirable, and to make sense of the prodigious might and moral ambiguity of western newcomers and their modernist agenda (Borofsky, 1987; Crocombe, 1994; Flinn, 1990; Howard, 1990; Lawson, 1990, 1993; Linnekin, 1983; McArthur, 2000; Neumann, 1992; Sahlins, 1981, 1985, 2005; Silverman, 1971; Thomas, 1992a, 1992b, 1997; Toren, 1999; for works outside of the Pacific, see for instance Camino, 1994; Jackson, 1994; Wilk, 1995).

Narratives of traditionalist decline are particularly powerful in the contemporary Pacific (Keesing, 2000). The ethnographic literature on these narratives exposes a point of fundamental importance: that the discourse is not mere "nostalgia," a rose-tinted recollection of one's childhood. The lost utopia of custom that is being described is usually one far beyond the recollection of the people describing it, and the belief in decline is often as vigorous among the young as among the elderly (Tomlinson, 2004: 656). As Matt Tomlinson aptly puts it, it is not nostalgia but *lament* (Tomlinson, 2009: 10). Islanders invoke the idyllic past "not [merely] as the better days whose passing they nostalgically mourn, but as an alternative present for which they yearn and for which they might fight" (Neumann, 1992: 246). Even among the Telefolmin (Jorgensen, 1981), for whom dissipation is the natural and unchangeable way of the universe, decline is a cause for action—indeed, the central idea around which action is organized. As Jorgensen writes, "Entropy is not merely contemplated, it is experienced . . . Order . . . is not merely an aesthetic preoccupation but a vital necessity—it is the means whereby men achieve some measure of control over their own lives" (Jorgensen, 1981:272). Similarly, for a Kwaio traditionalist activist, custom must be preserved not for decorative or sentimental reasons but as a way of "preserving the fabric of social life" (Keesing, 1994: 194). Narratives of decline, then, are not simply a grumpy conservatism among the elderly or dispirited reflections on colonial disempowerment but lively moral cosmologies that can inspire action.

The fact that I focus on a single, more or less agreed-upon preexisting worldview and a single, more or less agreed-upon set of attitudes toward climate change shows that I am bucking a particular trend in contemporary anthropology. I depart from the postmodern anthropological emphasis on heterogeneity, contestation, and context-specificity in society and the related critique, from postcultural anthropology, of "culture" as an inherently flawed concept due to assumptions of uniformity and boundedness. In Chapter 1, after certain ethnographic material has been presented, I will justify this perhaps unconventional approach to homogeneity/heterogeneity. But for the unconvinced reader, I will also justify it here on a priori grounds. While it is true that homogeneity and context-independence are risky assumptions to make, it is also true that heterogeneity and context-dependence are assumptions too: they can no more be safely assumed from the get-go than the latter and run an equal risk of misrepresenting the situation. Although an approach emphasizing homogeneity and context-independence of course risks essentialism and oversimplification, it is also true that the opposite emphasis risks reducing people's deeply held convictions to opportunistic whims and superficial epiphenomena—an implication indeed verged upon by some scholars (see Tyler [1986] for an extreme example; Carucci [1998: 15] for a more measured one). In the Marshall Islands, significant (though not total) agreement on certain attitudes to climate change does exist: and this means that Marshall Islanders have reasons to adopt certain understandings of, and responses to, climate change irrespective of context. So reasons must be found that do not hinge on momentary social situations or rhetorical needs. If, for instance, basically alike approaches to climate change blame and responsibility can be seen in public statements, private statements, interviews, and actions among a wide swathe of Marshall Islanders young and old (Chapter 4), then there must be a reason deeper and more enduring than the quickly shifting sands of moment-to-moment context. The utility of this approach has been demonstrated in such works as Mortreux and Barnett (2009) and Farbotko and Lazrus (2012) on Tuvaluan climate change attitudes and Jasanoff (2005) on the reception of biotechnology in the US, UK, and Germany: these works present a society's dominant attitude toward a particular scientific issue rather than emphasizing contestation and inconsistency. In this consensus-focused approach, the worst sins of representation can be avoided: essentialism can be avoided by historicizing the dominant attitude, reductionism by exposing its complexity and nuance, and homogenization by acknowledging but not overemphasizing discordant voices and inconsistencies within worldviews.

I also depart at times from the characterization of "modernity" and "tradition" as unstable, processual, and potentially unifiable entities (Borofsky, 1987; Jackson, 1994; Toren, 1999). There is of course considerable truth to these assertions: the Enlightenment understanding of modernity and tradition as distinct and irresolvably opposed is unsatisfying, and in Chapter 1 I analyze Marshallese notions of "tradition" and "modernity" as historically produced ideas rather than fixed truths. But again, an emphasis on instability can obscure as much as it reveals. In certain societies at particular times in history, fairly well defined and relatively stable characterizations of (what we can label) "tradition" and "modernity" may exist. Moreover, these concepts, however recent and variable they may seem to an outside observer, are often passionately asserted by insiders to be ancient and unchanging. Stability is often a core feature of the concept of "tradition," and it is this sense of stalwart staying power that lends it its rhetorical weight. Furthermore, even if the society in question epitomizes for the academic observer the potential unifiability of tradition and modernity, for insiders the two may be understood as irreconcilably opposed, and they may derive their symbolic power precisely from this contrast. As I will explore in Chapter 1, this in indeed the discursive situation in the Marshall Islands. In this book I thus take the mutability of tradition, the lack of a clear distinction between it and modernity, as etic but not emic truths.

METHODOLOGY

This book is based on one year and nine months spent in the Marshall Islands. In 2003 to 2004, working under the auspices of the NGO WorldTeach, I lived for nine months on Ujae, a remote and traditionally oriented outer island, and two months in the capital city of Majuro, as an teacher of English as a second language, Marshallese language textbook writer, and informal ethnographer. I returned for ethnographic fieldwork in July 2007, spending half of the summer on Ujae and half in Majuro, and again from May to September 2009, based primarily in Majuro but with some short expeditions to the outer islands of Leb, Mejit, Ailinglaplap, and Jaluit. I spent another month in Majuro in July 2012. Ranging from the copra sheds of remote islets to the air-conditioned hallways of Majuro's government ministries, I had opportunities to speak and spend time with Marshall Islanders from all walks of life. I employed a mixed-methods approach, including surveys, semistructured interviews, informal conversations, participant observation, and media analysis. In August and September 2009 I completed a survey of 146 adults in Majuro[4]; I also designed, with local climate change researcher and activist Mark Stege, a survey taken by ninety-four students at Marshall Islands High School, which was administered on paper by Stege in February 2009.[5] I conducted numerous semistructured interviews, recording about fifty hours of interviews in 2007 and 2009, in addition to many unrecorded interviews and countless informal conversations. Participant observation included my extended stint as the only foreigner on Ujae (2003–2004 and 2007), my attendance and presentation at various climate change-themed educational activities (2009), my involvement as interpreter and cultural liaison in a climatological expedition to the outer islands (2009), and many informal and everyday activities during my four stays in the country. Through all aspects of my fieldwork I used the Marshallese language without an interpreter except when speaking to certain well-educated informants who preferred to speak in English. Some notes on the use of verbal data are presented in a footnote.[6]

Some limitations need to be acknowledged. Most of the data comes from Majuro, the capital city. While I conducted a number of interviews on certain outer islands—Ujae, Ailinglaplap, Leb, Mejit, and Jaluit—I was not able to do so on other outer islands or in the city of Ebeye, and my survey results come entirely from Majuro. Also, since educational activities occur predominantly in Majuro as opposed to other communities, all of my participant observation at such activities was conducted in that city. This book may thus represent a somewhat Majuro-centric view of Marshallese climate change attitudes, but nothing I encountered or heard of in the outer islands leads me to believe that this badly misrepresents the country as a whole.

Other limitations of this study are by design rather than necessity. This book is not about the Marshallese government's response to climate change, such as its efforts to broadcast the country's plight to the international media and to campaign against climate change through multinational organizations. This book is not, at heart, about Marshallese ecological knowledge. Such an "ecological" focus would invoke the same western compartmentalization of climate change as "environmental" that I earlier critiqued. It would also ignore the voices of locals, who quite understandably do not think of the looming destruction of their country as strictly environmental in the usual western sense. Indeed one of this book's central arguments is that for most Marshall Islanders, climate change is not an "environmental" threat at all. While my research could be considered part of the anthropology of disaster, it is not typical of it: as the catastrophe has yet to occur in the Marshall Islands (notwithstanding some moderate impacts and frightening omens), this study focuses on the expectation of disaster, the "anticipatory dimensions" (Lahsen, 2007a: 9) of climate change, the "not-yet-event as stimulus to action" (Beck, 1992: 33). To put the same point differently, this study focuses on a different sort of climate change adaptation and resilience: not physical adaptation to climatic impacts, but mental adaptation to climatic predictions; not societal resilience to a present threat, but ideological resilience to a looming one. This topic is not a poor cousin to "real" adaptation studies: as a team of psychologists observe, "Social representations of environmental threats can themselves have dramatic psychosocial impacts" (Swim et al., 2009: 35), and nowhere is this more likely to be true than in a country where the environmental threat in question implies nothing less than forever losing the homeland.

1 Modernity the Trickster
From First Contact to the Postcolonial State

> It seems like life has tripped by at this gentle pace for centuries, and will continue to do so for centuries more. . . . But change has come to Tuvalu. (Lynas, 2004: 81)

Thus a climate change journalist describes an iconic low-lying nation destined, we fear, for a watery grave. The narrative is familiar: the world intrudes upon an untouched people; never before exposed to change, they are devastated by the encounter. The story has changed only in its referents: the rising sea, rather than the Bible or the Coke bottle, now plays the part of civilization's first taint. Unsurprisingly, the "first contact" theory of indigenous climate change attitudes does not hold in the Marshall Islands. More frontline than backwater, this nation has experienced missionization and imperialism, history's largest war and humanity's most destructive technology, at first hand and much of it within living memory. Marshall Islanders, we will find, share the aforementioned journalist's belief in a traditional paradise lost but not his assumption of a recent breaking point: for locals the decline began long before climate change, and today's rising oceans and temperatures are merely the latest in a string of corrupting encounters with the agents of modernity. To understand climate change from local viewpoints, therefore, we must examine the way in which a long history of contact has shaped Marshallese views of the local and the foreign, tradition and modernity, and by extension their understandings of modernity's climatic final act.

A BRIEF HISTORY OF CONTACT

Austronesian voyagers discovered what we now call the Marshall Islands at least two thousand years ago (Weisler, 2001a: 3). For the first settlers life could not have been easy. The atolls were poorly endowed in fertile soils, fresh water, and land area: they comprised 181 square kilometers spread between 1,225 individual islets. During typhoons, the narrow, flat islets could be quickly denuded of vegetation, cut in half, or even washed out of existence entirely (Erdland, 1961[1914]: 17–18; Kramer & Nevermann, 1938: 23–24; Spennemann, 1996). While the reefs offered more abundant resources, even these were vulnerable to depletion (Weisler, 1999). The shortcomings of atolls have led early explorers (Chamisso, 1986[1821]) and modern scholars (Sahlins, 1958: 218–246; Weisler, 2001b) alike to synonymize them with their deficiencies, characterizing them as "poor and dangerous reefs" (Chamisso, 1986[1821]) existing "on the margins of sustainability" (Weisler, 2001a); today they are synonymous with climate change vulnerability (Farbotko, 2005). But the feasibility of life on coral atolls is a result not only of environmental preconditions but also of human agency (Alkire, 1999; Thomas, 2009). Marshall Islanders preferentially settled on lagoon shores of leeward islets (Spennemann, 1996), areas that could remain relatively secure from storms over many centuries (Yamaguchi et al., 2005). They favored larger islets and more southerly atolls for their greater abundance of fresh water (Williamson & Sabath, 1982). While to the untrained eye much of the island landscape appears primeval, it is not so: the pioneers quickly transformed the wild islands into cultivated landscapes to fit their subsistence needs. The existence of staple crops—coconuts, pandanus, breadfruit, taro, and arrowroot—was no happy coincidence: the voyagers most likely brought these key species with them (Merlin et al., 1997). They refined their already well-developed canoe-building and navigational skills to levels perhaps surpassing even those of the Polynesians (Haddon & Hornell, 1975: 372), allowing them to maintain interatoll ties, which proved vital in times of need.

Thus the Marshallese managed not only to survive but indeed to produce a modest surplus, allowing for the development of stratification (see Sahlins, 1958). Chiefs (*irooj*) formed the political backbone of the islands. They had considerable authority over the land and demanded a share of its produce from the commoners (*kajoor*, literally "strength") in the form of first-fruits tribute (*ekkan* or *eọjek*) as well as their assistance in public works, defense, and conquest. In return a chief was expected to defend his people from invasion, provide for the victims of storms and droughts, and grant land rights to the deserving (Carucci, 1997a: 200; Poyer, 1997: 24)[1]: the Marshallese language contains a plethora of specialized terms for land parcels given by chiefs to commoners for specific reasons; *kwōdaelim*, for instance, is land granted to a commoner for having bailed the chief's canoe during a war voyage (see Kabua, 1993: 8–9). Other hierarchical relationships—parents above children, older siblings above younger siblings, *aḷap*s (lineage heads) above all others with rights on a land parcel—followed the same principle of unequal authority combined with mutual dependence and obligation: high people must care for (*lale*) low people, while low people must respect (*kautiej*, literally "make high") high people. The pre-Christian religion operated on similar principles: a small pantheon of gods (*ekajab*) from the spirit island of Eb in the west, as well as other ancestral and nonancestral spirits (*anij*), were supplicated with real or symbolic offerings of food in the hopes that they would reciprocate in the form of fair weather and good harvests (Erdland, 1961[1914]: 316–319).

Marshallese society, like so many others, was glued together with exchange. Distributing the fruits of the land cemented not only chief-commoner and spirit-human relationships but also kinship ties: puberty ceremonies and the

Keemem, a child's first birthday celebration, involved large-scale redistribution of produce and pandanus-leaf mats. Many Marshallese proverbs (*jabōn kōnnaan*) point to the social importance of exchange, in particular of food: *Ajej dikdik, kōjatdikdik* ("Sharing a little bit, hope") means "Give what you can, even if it is small." *Enrā bwe jen lale rere* ("Food basket so that we take care of each other") expresses the link between food and conviviality. *Etetal m̧ōm̧ōñāñā* ("Walking around, eating again and again") refers to the possibility of being fed again and again, at every house where one stops, in a properly generous Marshallese community. *Jab ālkwōj pein ak* (literally "Don't bend the frigate bird's wing") warns against disobeying those above you, especially the chief, but can also mean that one should never refuse an offer of food because the exchange itself is more important than the satisfaction of one's hunger. The two meanings are actually one: the proverb warns against neglecting one's duties in relationships of mutual dependence, in which one has an obligation to both give and receive.

With land as the preeminent form of wealth, competition over islands and adjoining reefs was the mainstay of traditional politics and warfare (see Kiste, 1974). An ambitious chief, or a particularly magnanimous one, might succeed in extending his hegemony across several atolls, but to our knowledge no chief ever conquered the entire archipelago (Mason, 1987: 11); the islands remained a mishmash of shifting chiefly domains. Probably for this reason, precolonial Marshall Islanders seem to have had little or no sense of themselves as a single unified people (Meto, 2000: 4) despite interatoll ties and recognized cultural and linguistic uniformity. They had names for the two chains of atolls that make up the country (Rālik and Ratak), toponyms that also easily became informal ethnonyms (Spennemann, 2000). But they had no term for the archipelago as a whole other than the vague *Aelōñ Kein* ("these islands") and no name for themselves as a people (Kramer & Nevermann, 1938: 13).

Such a unified identity would have to wait for the shared experience of colonialism. But the West largely ignored the islands until well into the nineteenth century. Spanish voyagers had occasionally visited as early as 1526 and had even claimed the islands as part of Spain's dominion. But the ownership was on paper only: the Spanish were far more interested in the Spice Islands than in these economically insignificant reefs, and Spain's influence on the islands was limited to a handful of brief, forgotten landings. British ships rediscovered the islands in the eighteenth century, occasionally trading with the locals, but again the archipelago was only a pit stop on the way to more lucrative shores, in this case China's. A voyage headed by Captain John Marshall lent the islands the name that would later be adopted as a toponym (*M̧ajeļ*) and an ethnonym (*rim̧ajeļ*) by the people themselves.

More significant was the several-month-long stay of Russian explorer Otto von Kotzebue in 1816–1817, easily the most extensive interaction between Marshall Islanders and Europeans to have yet occurred and our first source of ethnographic information (see Chamisso, 1986[1821]; Kotzebue, 1821). Kotzebue quickly established friendly relations with the local chief by giving abundantly of iron tools, and he proved his might by firing the ship's cannon to demand the return of a captive sailor. Having shown both power and generosity, Kotzebue was declared a chief (Kotzebue, 1821: 101)—a typical pattern in these early Pacific Islander-European encounters and best understood as an exercise of Marshallese agency to use these newcomers to local advantage (Walsh, 2003). Leaving the islands, the ship's naturalist-poet Adelbert von Chamisso, having arrived with a Rousseauesque view of these "kind-hearted children of nature" (Kotzebue, 1821: 57) and having experienced nothing to challenge that view, wrote "The poor and dangerous reefs of [the Marshall Islands] have nothing that could attract Europeans, and we wish our . . . friends the good luck of remaining in their remoteness." (Chamisso, 1986[1821]: 233).

Chamisso's proclamation—the first clause, in any case—proved remarkably unprophetic. As the century progressed, European interest in the islands increased manyfold. First the Marshalls attracted whalers and blackbirders[2], seeking to plunder the region's cetaceous and human resources. The abuses of these profiteers turned the formerly friendly islanders hostile, and the archipelago acquired a reputation for danger (Hezel, 1983: 200). Perhaps more encouraged than daunted by this ill repute, the first missionaries chose this time to arrive. They were Americans, a pair of Congregationalists sponsored by the New England-based American Board of Commissioners for Foreign Missions. Arriving on Ebon Atoll in 1857, they received a cordial welcome from the powerful chief Kaibuke (Hezel, 1983: 201; Utter, 1999: 34–35), who was evidently eager for a profitable alliance. Without local opposition, missionary activity proceeded smoothly: the missionaries quickly set about establishing a school and church, devising a written form of the language, translating parts of the Bible, and preaching the gospel to a growing congregation. Locals increasingly observed the Sabbath, abandoned non-Christian rituals, and became more secretive about tattooing, given its spiritual associations (Hezel, 1983: 205; Spennemann, 1993: 72). Other Protestant proselytizers, both Americans and Hawaiians, began to arrive in the islands. Initially, the missionaries had been able to secure the local chief's goodwill by offering him gifts and promising alliance. But soon the influence of Christianity—and the prospect for any Marshall Islander, even a commoner, to obtain wealth and prestige through cooperation with the missionaries—had begun to undermine chiefly authority. Kaibuke launched a backlash, but it was now too late for effective resistance (Hezel, 1983: 208–210). A mere fifteen years after the first missionaries arrived, islanders had become their own missionaries, spreading the word unaided and heading most of the country's churches themselves (Hezel, 1983: 210).

Conversion to Christianity marked the first major foreign influence upon the islands, and it facilitated further social change. But the islands nonetheless remained politically independent and unconquered. This changed in 1884, when Germany, lured by an already burgeoning trade in copra (coconut meat), bought the territory from Spain. German rule was the islanders' first real taste of imperialism, although the Germans governed with a light touch. Rather than challenging the authority of the chiefs, the Germans made them intermediaries in governance and trade, enriching them and arguably entrenching their power in the process (Walsh, 2003: 173). They outlawed the archipelago's chronic warfare (Spoehr, 1949: 90), thus stabilizing the country if also rendering inflexible the social system and distribution of resources. They introduced abundant trade goods and promoted copra cash cropping. Subsistence was not abandoned but was scaled back as imported food purchased with cash began to supplant home-grown produce and as taro and breadfruit land was replanted with coconut trees, creating the landscape that one still sees today.

In 1914, German commercial interest in the islands having waned and matters closer to home having become rather more pressing, Japan took unofficial control of the islands. That occupation became official after World War I, when the new League of Nations granted the territory to Japan as a mandate. In many ways the Japanese ruled more heavy-handedly than the Germans: they weakened rather than strengthened the chiefs (Walsh, 2003: 179–189), removed the chiefs' exclusive ownership of particular tabooed reefs (Tobin, 1952: 11), and even tried (unsuccessfully) to replace the islanders' predominantly matrilineal inheritance patterns with a Japanese-modeled patrilineal system. Nonetheless, elderly Marshallese remember the Japanese time as agreeable. But it would not last. In the run-up to the Pacific War, Japan withdrew from the League of Nations, imported military personnel to the island, and heavily fortified many atolls in this strategic eastern edge of its Pacific empire. The archipelago witnessed heavy fighting in the war. As bombs fell, supply lines failed, food stocks dwindled, the Japanese soldiers became desperate and cruel to the Marshallese—forcing them to work, confiscating their food, and banning Christian activities (Poyer, 1997: 31; Spoehr, 1949: 33).

It is therefore with fondness that elderly Marshallese now recall the victorious arrival of US forces. As the Americans expelled the Japanese, they ingratiated themselves to the locals, giving food freely and reinstating church services. The Marshalls had changed colonial hands once again, becoming an American-administered United Nations Trust Territory—a de facto American possession. America was now the islanders' third colonial master in little more than three decades. Again the remote, resource-poor archipelago, presumably peripheral to world affairs, with "nothing to attract Europeans," would be at the forefront of global affairs. The United States chose the territory as its premier nuclear testing site. Relocating the people of Bikini and Eniwetok Atolls in the northern Marshalls, the American military conducted sixty-seven nuclear tests through the 1940s and 1950s. The tragic saga of relocation, irradiation, coverup, legal confrontation, and compensation continues to the present day.

With the testing concluded, the Marshallese leadership hoped for self-governance yet recognized that the islands were now highly dependent on American subsidies and would likely need to remain so for many decades. The desire for political independence and the reality of financial dependence led to a compromise in 1986 with the Compact of Free Association, a treaty between the United States and a newly sovereign Republic of the Marshall Islands (RMI). The compact makes the archipelago as close as possible to a US possession while not actually being one. That is, the country is self-governing and has its own seat in the United Nations, yet roughly two thirds of its budget derives from various forms of American aid (RMI Government, 1996: 3), its currency is the US dollar, it is eligible for US federal programs, its defense is a responsibility of the United States, and Marshallese citizens can serve in the US armed forces and freely live and work in the United States. In addition, the United States maintains a military base on Kwajalein Atoll, part of the Ronald Reagan Ballistic Missile Defense Test Site, where antimissile defense technology is tested; those with traditional claims to the atoll have been enriched by rent payments. The RMI and the US remain as closely tied as any two countries can be.

In the 1990s, another foreign group began to exert influence on Marshallese society. Through a government-run passport selling scheme, a few hundred mainland Chinese acquired Marshallese citizenship, settled in Majuro, and opened shops that outperformed local businesses. While only 2% of citizens are nonindigenous (RMI Government, 2005: 20), the existence of foreigners, in particular Chinese, looms large for Marshall Islanders and is a subject of songs, newspaper articles, and political cartoons.

The Marshallese population is itself in considerable flux. Urbanization has proceeded rapidly since World War II (Juumemmej, 2006: 20). Two thirds of Marshall Islanders (Juumemmej, 2006: 73) now live in the country's two urban centers—Majuro, the capital and largest city, with around 25,000 inhabitants, and Ebeye, population 11,000, a bedroom community in Kwajalein Atoll for Marshallese who work at the nearby US military base. The rest make their homes in the hundreds of small villages scattered over the country's twenty-nine atolls and five single islands, where subsistence and copra farming are economic mainstays, amenities like running water and electricity are scarce, and trade goods may be available only sporadically. The majority of the population therefore lives on land leased from traditional landowners rather than on their own heritage land, depending primarily on cash rather than subsistence. Extremely high birth rates since World War II have skewed the demographics toward children and young adults (Juumemmej, 2006: 79) and precipitated a population boom. Since this has reached nearly the point of saturation, large numbers of Marshall Islanders, confronted with few economic opportunities at home, have migrated to America. The years 2000–2002 alone saw 5,000 Marshallese, or a tenth of the country's population, leave for the United States (Pacific Institute of Public Policy, 2010: 3), settling in particular in Hawaii, California, Oregon, Washington, Arkansas, and Oklahoma. By 2010, more than 22,000 Marshall Islanders—more than a quarter of the total population—were living in the US (United States Census Bureau, 2012), the number having tripled since a decade before (United States Census Bureau, 2012: 16). The older religious denominations (the United Church of Christ, the continuation of the nineteenth century's Protestant missions, with 51.5% of the current population; and Catholics, from a second wave of missionization in the 1890s, with 8.4% of the current population) now rub shoulders with Assemblies of God (a Pentecostal denomination, 24.2%), Bukot Non Jesus ("Searching for Jesus," another Pentecostal denomination, 2.2%), Mormons (8.3% and growing), and small numbers of Seventh-Day Adventists, Baha'i, and Jehovah's Witnesses (Freedom Report, 2009).

THE MARSHALLESE WAY

Much of present-day Marshallese discourse can be understood as attempts to come to terms, intellectually and emotionally, with this history of foreign contact and social change. And much of Marshall Islanders' sense of who they are as a people—even the fact that they consider themselves a people—finds its origins in colonial encounters.

Through all of the regimes—German, Japanese, American, and Marshallese—the political unification of the islands under a single administration encouraged a feeling of unified identity. So too did the many challenges to precontact Marshallese lifeways. The missionaries had little interest in or appreciation for traditional religion: they called upon locals to renounce their evil customs rather than incorporate them into Christianity (Erdland, 1961[1914]: 305; Utter, 1999: 36–37). The Germans outlawed warfare, weakened subsistence, and enriched chiefs to the extent that they no longer needed to ingratiate themselves to the commoners. The Japanese challenged the supremacy of the chiefs and sought to eradicate matriliny. American nuclear testing shattered traditional livelihoods on several atolls. Chinese immigration threatened the cultural homogeneity of the homeland and the viability of Marshallese-run businesses in Majuro. Outmigration to the US immersed thousands of Marshall Islanders in such foreign cities as Springdale, Honolulu, and Spokane, forcing a conscious choice between retaining Marshallese ways and abandoning them. The result in each case was a partial or total opposition between a unified "Marshallese way," prevailing in the past, and a foreign way, prevailing in the present—fertile ground for the invention of tradition. Marshall Islanders now had reason to "substantivize" (Thomas, 1992a) their precolonial social system, reifying it as an object for conscious consideration, even as they abandoned parts of it and saw other parts stolen.

A unified ethnic identity is now vigorous in the Marshall Islands.[3] People strongly and effusively identify as Marshall Islanders (*riṃajeḷ* or *armej in ṃajeḷ*), in distinction to other ethnic groups such as Americans (*ripālle*) and Chinese (*rijāina*). They speak often and passionately of their tradition (*ṃanit*) and the Marshallese way of life (*ṃantin ṃajeḷ*), contrasting it with other peoples' lifeways, in particular American culture (*ṃantin pālle*). Marshallese custom, it is frequently said, is the one correct way for

Marshall Islanders to live and the greatest way of life in the world. One of the best compliments a foreigner in the Marshall Islands can receive is to be told (not literally but nonetheless approvingly) that he or she is Marshallese. Publicly criticizing *ṃanit* is close to unthinkable; only well-educated individuals will do so, and only in private. The Marshallese constitution declares that tradition is "a sacred heritage which we pledge ourselves to maintain," and that sacred heritage is celebrated self-consciously in yearly Manit Day festivities.

To be sure, Marshall Islanders sometimes disagree on what exactly it means to be a *riṃajeḷ* and to follow *ṃantin ṃajeḷ*. But consensus predominates. Certain elements are repeatedly emphasized by a wide range of citizens. First is subsistence, indicated in the Marshallese language by the phrase *ejjeḷọk wōṇāān* ("free of charge"). The informal Marshallese name for the nation, *aelōñ kein* ("these islands"), is said to refer to the three zones from which people eat: *ae* ("ocean current") representing the ocean, *lōñ* ("up") representing the sky, and *kein*, representing the land. Marshall Islanders, it is said, eat from all three spheres—plants from the land, fish from the sea, and birds from the sky—and such a subsistence lifestyle is said to guarantee an easy abundance for all; freeness is freedom. Marshallese foods (*ṃōñā in ṃajeḷ*)—in particular coconuts, breadfruit, and pandanus—are lauded often and enthusiastically. Their enjoyment by a foreigner is spoken of as a prime indicator of going native, and their refusal is an insult to the hosts and a symbolic rejection of Marshallese culture in general.

Second is conviviality: kindness and generosity within and between families (*ippān doon*, "togetherness"; *lale doon*, "taking care of one another"; or *iọkwe doon*, "loving one another") as well as proper respect (*kautiej*) for those in hierarchical roles above oneself—such as parents, older siblings, and chiefs. Also included here is appropriate support (*lale*) for those below oneself—such as children, younger siblings, and commoners. These proper and generous relations are often said to be the bedrock of *ṃantin ṃajeḷ*. Subsistence and conviviality are closely connected; indeed, for Marshall Islanders they are nearly synonymous: subsistence promotes conviviality because when food is free of charge, people will generously share it, fostering togetherness, it is said. (For more on this discourse, see Rudiak-Gould, 2009a: 17–39).

Third is chiefs and land. Respect for the chiefly establishment is often said to be the linchpin of Marshallese tradition. Land (*bwidej*) is also highlighted. The Marshallese constitution declares that the people "valu[e] nothing more dearly than our rightful home on these islands." Well-educated locals, in particular, state in so many words that Marshallese identity depends upon the land. The archipelago (*aelōñ kein*, "these islands"; *aelōñ kein ad*, "our islands"; *aelōñ in ṃajeḷ* or *ṃajeḷ*, "the Marshall Islands") is said to be a gift from God (*mennin leḷọk jān Anij*) and a precious inheritance (*jolōt*) from the ancestors: there is no other homeland for the Marshallese, and the Marshall Islands is homeland to no other. Most Marshall

Islanders have no knowledge of or interest in prehistoric Pacific migrations; although they consider themselves related to other Pacific Islanders, they never speak of their ancestors having originated in Southeast Asia or having arrived in the Marshall Islands by way of other archipelagos. As far as they are concerned, their people have always lived in the Marshall Islands and have always been Marshallese. Transplants to urban centers still identify as people of their home atolls even when they have no intention of returning, for it is there that they have their *ļāṃoran* (or *kapijukunen*): their heritage land, which has belonged to their matrilineage (*bwij*) for many generations.[4] Marshallese immigrants to the United States remember their Marshallese origins and their home atolls, in many cases still occasionally paying tribute to the chief under whose authority their heritage land falls.

The connections between these emblems of Marshallese identity are intimate: subsistence fosters conviviality, conviviality requires hierarchy, hierarchy is invested in chiefship, chiefly authority depends on land, and land grants subsistence. This is the good life, said to guarantee harmony and prosperity for all. Marshallese traditionalism is well expressed in a small but vigorous artistic movement that has recently arisen in Majuro (Journal, 2000, 2003c; Madsen, 2006), producing several dozen painted murals, including a very large piece now gracing the entire length of the outer wall of Marshall Islands High School in Majuro. In these works of art, life is as it should be. People are living off the land, presented as bountiful and perfectly maintained. Copious coconut-frond baskets are visible, overflowing with food, ready to be shared. The houses are thatched with local materials. People are graciously engaged in collective endeavors: house building, a *Keemem* ceremony, or the *alele* fishing method in which dozens of men surround a school of fish with a coconut-frond scarer. When alone, people are working to serve others: for men, the properly masculine duties of fetching and husking coconuts, fishing, sailing, and building canoes; for women, the properly feminine duties of preparing food, tending fires, taking care of children, and weaving pandanus-leaf mats. People rely entirely on local resources and customs, so life is proper, prosperous, and harmonious. This, in visual form, is the Marshallese discourse of traditional harmony.

Nicholas Thomas has argued (Thomas, 1992a,b, 1997) that the "tradition" that Pacific societies have come to reify is neither a random collection of precontact practices nor an invention from thin air nor a simple catalogue of what is "objectively" most important about the society, but instead comprises the specific cultural practices that colonialist and modernist discourse "subordinates and denigrates" (Thomas, 1997: 190). Thomas's theory applies well to the Marshall Islands. The cultural elements that Marshall Islanders emblematize—subsistence, conviviality, hierarchy, chiefship, and land—are precisely those that historical forces have threatened. In the era of missionization, as we saw previously, the power of foreigners rivaled that of chiefs even as the chiefs attempted to harness that foreign power to entrench their own hegemony. During the German administration, the advent of money challenged the precolonial system in the ways suggested above. The conversion of much of the Marshall Islands into copra plantations weakened the subsistence-based unmoneyed lifestyle. More fundamentally, money offered a new and subversive source of sustenance: with cash, one could have no stake in the land and no loyalty to a chief yet still be able to eat. Money supplanted land and buying supplanted growing: the chiefs' supremacy, and the entire social system that accompanies it, were powerfully threatened in symbolic terms, even if the chiefs' stake in the copra trade allowed them to maintain literal supremacy. Cash endangered land in a more direct sense as well: both German administrators in the early twentieth century and American administrators in the mid-twentieth century attempted to buy land outright from its traditional owners. In 1911, a German administrator offered chief Leit 18,600 marks for Rongerik, Rongelap, Ailinginae, Bikini, and Wotho atolls, to be used as copra plantations (Walsh, 2003: 174–176). In 1957, the American administration offered Amata Kabua—a powerful chief, landowner, and later the republic's first president—US$187,500 for indefinite use of 250 acres, effectively buying the land; at one point he was offered US$300,000 in cash on a table in front of him (Walsh, 2003: 222). In both case the offers were rejected and the land remained in the chiefs' hands, but the temptation and the threat were real enough: Chief Leit was interested in doubling the sum for Bikini alone before deciding to refuse the offer altogether.

American atomic testing mounted a further attack on the supremacy of land, chiefs, and subsistence. In some cases land was destroyed outright—three islets in Bikini Atoll were vaporized, leaving nothing but submerged craters in the reef—while in other cases the land remained but lingering radiation robbed islanders of their ability to enjoy its fruits. Evacuated to other islands, the nuclear refugees found themselves on inferior land, uninhabited for a reason, where subsistence was difficult or impossible (Dibblin, 1988; Kiste, 1974, 1977; Niedenthal, 2001). In the most famous case, Bikinians were relocated to Rongerik Atoll, where they nearly starved owing to the poisonousness of many of the fish, and then to Kili Island, the "prison island" which had been only sparsely and sporadically inhabited in precolonial times because of its poor anchorage and lack of a lagoon (Kiste, 1974, 1977; Niedenthal, 2001). Some later moved to Ejit Island in Majuro Atoll, an islet far too small for farming. In these and other cases of nuclear exodus, men, unable to provide, lost their "moral standing and self-esteem" (Barker, 2004: 75), and the traditional leadership was undermined because its power had depended on the control of productive land (Barker, 2004: 75–77). Land lost, the people were compensated in *money*, in the form of charity and reparations. Thus the affected communities had been forced to sell their islands, to trade a subsistence lifestyle for a moneyed one.

Other threats have appeared more recently. The formal power of chiefs, if not their informal influence, has been reduced by a democratic system that invests legislative authority in an elected parliament, the Nitijela, and relegates chiefs to a council of Iroij with merely advisory capabilities; chiefly powers are now less enforceable, even if chiefs have retained much power by being elected to senator seats and appointed to the presidency. The progressivist United Democratic Party, opposed by the more self-consciously traditionalist Aelōñ Kein Ad Party, has suggested expropriating certain plots and making them public land, thus ensuring the smooth functioning of government services such as schools. The migration of the population into urban centers, far too densely populated for horticulture, as well as the movement of thousands of Marshallese to America, where moneyed life is inevitable, have also threatened subsistence; land-based chiefly authority has suffered from the resulting abandonment, in body if not in sentiment, of *ḷāṃoran*, heritage land. Meanwhile, an American has managed to purchase subordinate land rights but not chiefly rights to part of an islet in Majuro Atoll. In the Marshallese view, conviviality has come under siege by Chinese immigration; Chinese-run shops have displaced Marshallese businesses in part because Chinese immigrants, being outside of the Marshallese system of mandatory reciprocity, can administer their businesses like good acquisitive capitalists. In opposition to this, Marshall Islanders emphasize their communalism and ethic of sharing. Subsistence, conviviality, hierarchy, chiefship, and land have become the emblems of Marshallese culture because all are in the crosshairs of history.

An outsider might say that tradition and modernity, the local and the foreign, are not so distinct and incompatible, that Marshall Islanders have in fact successfully hybridized precontact practices with postcontact realities, creating a "Marshallese modernity." But Marshall Islanders themselves rarely speak of it this way. Marshallese tradition is for them defined by its fundamental, diametric difference from foreign and modern lifeways. Unsurprisingly, given the country's postwar history, it is Americans (*ripālle*) who most often play the conceptual role of the Other. American culture (*ṃantin pālle*) is spoken of as the polar opposite of *ṃantin ṃajeḷ*: subsistence is replaced by a cash economy (*mour kōn ṃani*, "living by money") and conviviality is replaced with individualism (*kwe wōt kwe, ña wōt ña*, "you are just you, I am just me"). Just as subsistence and conviviality are seen to mutually reinforce each other, so too are the cash economy and individualism: money is inherently divisive, something that only *some* people can possess, which leads inevitably to selfishness, discord, and division. Chinese people are also sometimes cast as the cultural Others. In contrast to the compassionate and peaceful ways of Marshall Islanders, Chinese people are said to be rude, aggressive, and greedy, caring only about money; they steal, disrespect Marshallese custom, and hurt local businesses. They are also dirty, smelly, purveyors of inferior goods, and the Chinese women are prostitutes. The negative statements about American culture are not applied to individual Americans, who are in fact welcomed warmly. The negative discourse about US citizens can be considered mainly a conceptual aid, good to think with. In contrast, the unfavorable stereotypes of Chinese are applied to individuals: there is real resentment here, not merely a discourse of alterity.

Perhaps the most fundamental attribute that Marshall Islanders associate with these foreigners, and repudiate, is money. "Living by money" (*mour kōn ṃani*) becomes a metonym for everything bad about modernity and everything antithetical to Marshallese custom. It is closely associated with America, a conceptual link strengthened by the country's use of the American dollar. Money is said to undermine the lifestyle of easy subsistence by requiring people to buy rather than grow and catch food. Money destroys conviviality as well. A song called *Ṃani ej okran nana*, "Money is the root of evil," declares (my translation): "It will be hard, so hard, to rid ourselves of this disease." Political cartoons in the *Marshall Islands Journal* often draw sad portraits of a society fractured by greed. In one cartoon,[5] a man asks his older brother for $20 so that he can buy rice for his family. The man's brother replies, "I'm sorry, but I don't know who you are." An onlooker exclaims, "Wow, money has really changed his thinking. That's your older brother!"

Money makes people forget their chiefs and their land, it is feared. The only form of infinitely storageable good in traditional Marshallese society was land (in the extended sense of islands and adjoining reefs), making it the preeminent form of wealth. With chiefs considered the ultimate authorities over all land, it was impossible to be entirely outside of their sway; more generally, with land being the only source of sustenance, one had to belong to the hierarchical system of tenure whether one wished to or not. With the advent of money, a new source of sustenance burst onto the scene. Easily stored, easily hidden, cash is uniquely suited to hoarding, allowing and encouraging greed. Moreover it is anonymous: unlike a hand-woven mat, it bears no traces of who made it; unlike a land parcel (*wāto*), it bears no marks of its previous owner. It is perfectly impersonal (Bloch & Parry, 1989: 6). All of this makes it, in the Marshallese view, the ideal carrier of individualism (*kwe wōt kwe, ña wōt ña*), something that Marshallese discourse reviles as antithetical to conviviality and hierarchy.

CULTURAL DECLINE AND SEDUCTIVE MODERNITY

The past (*jeṃaan*, *etto*, *raan ko ḷọk*) is considered the site of Marshallese tradition, a time when custom was upheld, chiefs were revered, conviviality was vigorous, strife and privation were unknown; it is described as a golden age resembling the paintings and murals I described earlier. But this arcadia, it is said, did not last. It fell prey to the advent of outsiders, especially Americans, whose foreign ways lured islanders away from their *ṃanit*. With the Americanization and foreignization of Marshallese society, it is said, the good life transformed into its degenerate mirror image: the contrast between past and present is identical to that between Marshallese culture and American culture. In countless ways Marshall Islanders nowadays violate tradition (*kọkurre ṃanit*), locals say. "In the future we won't have brown skin, we'll have white skin!" said one man, metaphorically, with a laugh. The Marshallese language (*kajin ṃajeḷ*) has declined, people say, and this decline is an indicator or proxy for the decline of *ṃantin ṃajeḷ* in general. Children—and even government officials—do not speak proper Marshallese any more: they use too many English words and they have forgotten *kajin etto*, the old Marshallese language. Soon Marshall Islanders will simply speak English, it is said. The language is considered an integral part of the culture, and its replacement with English a sign of the replacement of Marshallese culture with American culture. Meanwhile people decry the loss of traditional lore and skills, such as *katu* (weather forecasting) and *meto* (navigation). The influx of foreigners is seen as a cause, a symptom, and a symbol of tradition's demise. Marshall Islanders have learned the American way, in all its degeneracy, from Americans themselves, as well as from American films as vehicles of seductive foreign ideas. "Soon our children will be half American, half Chinese," it is said. Furthermore, people say, as these foreigners flood into the country, locals leave in droves for the United States.

Meanwhile, subsistence and home-grown produce are giving way to cash and foreign imports, locals lament. People these days are said to prefer imported staples such as rice, flour, and sugar over local crops such as coconut, breadfruit, pandanus, and taro, even though local crops are healthier, more conducive to conviviality, and more authentic to Marshallese culture, in addition to being free for the taking. People are said to be lazy and therefore to prefer Spam, canned tuna, and canned mackerel over seafood that they must catch themselves. Some crops, like arrowroot, are said not to grow well anymore because of modern changes such as radiation. A man on Leb expressed the difference between past abundance and present scarcity in the form of fruit: he pointed to a small, misshapen, discolored breadfruit and said, "You see this breadfruit? It is bad. It is a *mā in raan kein*—a breadfruit of nowadays." Then he showed me a plump, shapely, healthily green breadfruit and said, "And you see this one? It is good. It is a *mā in jeṃaan*—a breadfruit of the past." More generally people's relationship to the land is weakening. A Bikinian man lamented that the land on Bikini Atoll was degraded, not merely because of nuclear testing but because Bikinians were no longer there to tend to it. A well-educated woman in Majuro told me that her attachment to her heritage land was strong, but her daughter, heir to those same landholdings, had no interest in them because she had grown up in Hawaii; the pattern, said the woman, was typical.

As land is supplanted by money and abundance by scarcity, conviviality gives way to social strife. Whereas people once ate together at one fire, they now eat separately, family by family or individual by individual, it is said. People "grieve alone." They are rude and stingy toward those without money, even their own kin. They fail to give freely as they once did, lying to each other that they have nothing in order to avoid sharing.

They assault and kill each other and beat their children in ways previously unheard of. They violate kinship taboos, acting too freely around proscribed relatives, and they are remiss in their hierarchical obligations. In particular, low people fail to properly respect high people: children are disobedient and slothful, and commoners fail to present a food tribute to the chief. Women fail to be properly meek: they break the old rules by which a woman should not shout in the road, should not sit in high places or climb trees. They drink, smoke, flirt, and wear trousers, things that only men should do. They wear immodestly short skirts and sell their bodies for money. High people fail to properly take care of low people: men are lazy and feckless, chiefs care more about amassing money than feeding their commoners, parents abuse their children, and many children have no fathers. On all sides and in all ways, it is said, the harmony, love, generosity, respect, and propriety of the past are disintegrating into discord, hatred, meanness, insolence, and indecency.

The following statements collected in 2007—the first two from outer islanders of limited formal education, the last two from well-educated urbanites (specifically then-President Kessai Note and then-mayor Eldon Note of Kili, Bikini, and Ejit)—exemplify this discourse of decline:

> I think the Marshallese customs of the past are gone. In the past, they didn't say, "Oh, buy it. Buy fish. Buy coconuts." Everyone took care of each other. In the past, the old days, when a canoe broke and drifted from Majuro to Ujae, and they saw the castaways, they would say, "Oh!" The chief would say . . . "Take care of those people." That's tradition. . . . They would bring the castaways and put them with their kinsfolk. . . . And then they would take care of each other. They took care of each other very well. "Oh, hello! Come and eat."

> The past was good, because there was no airplane, [so] nothing came in from abroad. . . . When you made *bwiro* [preserved breadfruit] . . . everyone on the island came. Absolutely everyone. They came and brought food. . . . Everyone ate together. . . . The men carved canoes . . . and the women brought [food]. . . . Now this has disappeared. . . . The people of the past were very tall and large. But people these days are very short and thin, because things arrived that didn't previously exist in the Marshall Islands, like cigarettes, alcohol, rice, and flour. We use them so much. We used to eat only pandanus and Marshallese food. Now we're very sick with diabetes.

> There's no question that life was better in the past. Everyone worked together, helped each other, fed each other, and made food. No one was hungry. Everyone worked and ate and respected their chief. Nowadays, you see crime, theft and fighting. None of these things existed before because they were against Marshallese custom. They didn't exist because

> taboos and chiefs, elders in the community, and one's own family controlled people. They made sure their children and friends didn't do bad things or make trouble. No one stole and no one fought, because everyone was provided for. They weren't hungry. There wasn't any money or things like that yet. It's harder nowadays because of the economic reality. People need money for all kinds of things. Once westernization, western culture and all of those things came, people were influenced by it. They need money, and they need things that aren't in their community.

> Ever since we moved from Bikini . . . we lost, I might say, our culture. . . . Culture, custom, togetherness, respect [of] each other. . . . They are all gone. . . . The community on Bikini, they used to be a community of sharing things together. . . . It was like a family. . . . People were gathering and eating all night, never thinking about what they were going to eat tomorrow. . . . When you finished the meal, you would never think about the next meal. It was an easy life.

All of these troubles are said to have struck the urban centers far more severely than the outer islands, both urbanites and outer islanders agree. In the outer islands, "one smells only the scent of Marshallese culture" (*Kwōj āt wōt bwiin m̧antin m̧ajeļ*). When students in Majuro were asked to compare rural life with city life, all said they preferred the former (Journal, 2007b). A song called *Mour ilo Outer Island* ("Life in the outer islands"), composed in 2006 by the Rita Boys, celebrates its namesake (my translation): "Life in the outer islands is better than life in the city / Every day I make copra, every day I go fishing." A man from Ujae voiced a widespread sentiment when he told me,

> In Majuro you have to live entirely by money. You go to the hospital and you have to spend money. You go to church, and you spend money. We also buy canned food, and it's expensive. Nowadays gas is very expensive. There's a big difference between being in the urban centers and the outer islands, because in the outer islands you can eat Marshallese food. . . . On Ujae, you wake up, take a shower, and eat—eat native island food that we got from Marshallese trees. We don't buy it.

The urban centers, then, are to the outer islands as the past is to the present, and as the Marshall Islands as a whole is to America. The discourse is easily represented in binary fashion, with arrows showing the generally tragic flow of things (Figure 1.1).

Some Marshallese visual art aims to show the brutal contrast between those two poles. A mural painted on a bus shelter in Majuro is starkly divided into two sections. On the left is an island with coconut and breadfruit trees and no houses; the ground is green, a fish swims in the sea, and a man in traditional woven kilt, his long hair tied into a bun—a popular style

Eṃṃan mour	→	*Epen mour*
Good life		Difficult life
Jeṃaan	→	*Raan kein*
The past		The present
Ṃajeḷ	→	*Amedka*
The Marshall Islands		America
Outer island	→	*Center*
Outer islands		Urban centers
Ṃantin ṃajeḷ	→	*Ṃantin pālle*
Marshallese culture		American culture
Kajin ṃajeḷ	→	*Kajin pālle*
Marshallese language		English language
Lale doon; ippān doon	→	*Kwe wōt kwe, ña wōt ña*
Take care of each other; togetherness		Individualism
Kautiej	→	*Jab kautiej*
Respect (Hierarchy)		No respect (Lack of hierarchy)
Ejjeḷọk wōṇāān	→	*Mour kōn ṃani*
Free of charge (subsistence)		Live by money (cash economy)
Ṃōñā in ṃajeḷ	→	*Ṃōñā in pālle*
Marshallese food		American food

Figure 1.1 Binary analysis of the "modernity the trickster" discourse.

at the time of early western contact—fishes aboard a traditional *kōrkōr* canoe. On the right side is an island covered with modern buildings and bereft of trees; the ground is brown, a car drives on the road, and a factory spews smoke into the air and green filth into the ocean; two men in a motorboat are fishing with dynamite. It is no fluke that this depiction mixes and blurs what westerners would call degradations in "nature" and degradations in "culture," a point I will return to in Chapter 3.

If, in the Marshallese view, there is a single force that produces all of these ills, it is money. "Once the green paper came, it changed everything," an elderly man told me. A commentator in the *Marshall Islands Journal* (Murphy, 2007) bemoans that *ṃantin ṃajeḷ* ("the Marshallese way") is gone, replaced with *ṃantin ṃani* ("the way of money") and its violent results. People have taken the "t" out of *ṃanit* ("tradition") and turned it into *ṃani* ('money"), it is said. Locals cite *mour kōn ṃani* as the central malady of the present day; it could be translated literally as "living by money" or more freely as "the cash economy," but perhaps the most insightful translation is "modernity," because *mour kōn ṃani* refers to the entire lifeway of the present day. While recognizing the vagueness and polysemy of "modernity" and the imperfect fit between academic terms and local concepts, in this book I will nonetheless adopt the gloss of "modernity," as I believe it does little violence to the Marshallese concept.

All of these laments about the cash economy—"Pretty soon, money will be chief"—are understandable in terms of the attributes of cash previously discussed. People, one person said, used to serve the chief because he had land, but now they serve whoever has money. If the Bible says, "You cannot serve both God and money," Marshall Islanders would say, "You cannot serve both chief and money." For Marshall Islanders, modernity is bad because it provides security from an improper source: oneself and money rather than the chief and the land.

One of the most popular characters in Marshallese mythology, starring in numerous legends still told today, is Ļetao the trickster. One cannot talk of modernity without talking of Ļetao. He is much like the Polynesian character Maui (see Sinclair, 2001). He creates horrors disguised as wonders. He delights in violating taboos and crossing boundaries. He employs his cunning and magical prowess to swindle, humiliate, and kill. Like Maui and so many other trickster figures, he is the legendary bringer of fire: he impresses a boy by showing him how fire can be used to improve the taste of food, then tricks the boy by burning down his house (Williamson & Stone, 2001a: 88). He is also the bringer of sex (Carucci, 1986): necessary and dangerous, destructive and creative, just like fire. In traditional Marshallese mythology he is the grandson of the creator god Lowa (Erdland, 1961[1914]: 308–310; Williamson & Stone, 2001b: 32), while today he is associated with another powerful entity, namely the United States. At the end of Ļetao tales, the storyteller often says, "And then Ļetao sailed to America and lived there—and that's why Americans are so clever" (some add "but lie so much"). In this view, Americans (and by extension, money and modernity) are like Ļetao, clever and efficacious but also devious and subversive, giving something very different from that which they promise.

In one oft-told legend, Ļetao sails to one of the Gilbert Islands and meets a chief. He offers to make food for the chief and the chief's people by putting himself into a well-heated earth oven (*uṃ*). The Gilbertese chief balks, certain that the newcomer will be cooked to death in the process, but Ļetao insists. He confidently enters the steaming *uṃ*, covering himself with rocks. The chief and his people wait for hours. When they open the oven, they are amazed to find, in Ļetao's place, a cornucopia of delicious cooked food. Ļetao then reappears, triumphantly. The awestruck chief demands to try it himself. Ļetao warns him that the oven is very hot, but the chief is determined. The chief enters the oven, and his people cover him with stones and wait for several hours. When they open the oven, they find no food, only their well-done traditional leader. Furious, the people try to capture Ļetao, but he has already escaped.[6] He sails to America, and "that's why Americans are so clever" ("but lie so much").

What has happened symbolically in this narrative? When Ļetao produces his magical cornucopia of food, he demonstrates his ability to amply provide. He thus challenges the chief's authority, which derives from his ability to care for his subjects and therefore to demand their loyalty. Ļetao's magic threatens to supplant the chief's land as the source of food, wealth, life, and power, enticing the commoners to worship a new master. The chief must therefore learn the magic himself if he is to maintain his dominance. He dies in the attempt, a victim of the trickster's wiles. Now Ļetao's victory is complete. In a single cunning move, he has both killed the chief and proven him unable to provide. (Indeed these were the two avenues by which, traditionally, one chief could supplant another: by force or by superior guardianship.) Now the traditional system, the old way of securing the necessities of life, has been shattered: the people have lost the protection of their chief. Ļetao promised prosperity in exchange for sedition. They took the bait, only to find that they had destroyed the social system that supported them. When Ļetao escapes, they have lost, too, the prospect of his custody and the life-giving magic that comes with it; they are left with nothing. Tempted from the status quo by the prospect of an easier life, the islanders have lost even the modest prosperity that they began with.

This is exactly how Marshall Islanders speak of money, modernity, and technology: in terms of the power that Ļetao brings with him to America at the legend's end. They are, like Ļetao, seductive newcomers, flaunting their ability to provide more ably than land, chief, subsistence, and tradition. The newcomer's power and wealth are unquestionably prodigious, and the islanders are enticed. Yet when they agree to the deal, pledge their allegiance to the United States, they find that the same technology that brings the trickster such power brings local people only hardship and discord; *m̧ani*, apparently a greater provider than *m̧anit*, ends up making life harder, not easier. Ļetao's unceremonious departure echoes the looming discontinuation of American aid, which forms the cornerstone of the Marshallese economy: when the cash flow ceases, people will find that they have already jettisoned their traditions—baked their chief—and have nothing to live on; the old provider is dead and the new provider has run away. Thus, in the local view, Marshall Islanders have fallen for the same Faustian bargain that cooked the Gilbertese chief. Be tempted by a new way and you will live to regret it: this is the message and the fear expressed by the legend of Ļetao and the chief.[7]

Even the briefest summary of twentieth-century American intervention in the Marshall Islands makes the trickster image seem apt. The United States displayed a creative, destructive power worthy of a Ļetao, Maui, or Prometheus. For the people of Rongelap Atoll, March 1, 1954, was the "day of two suns," because the light of the nuclear explosion in the west equaled that of the rising sun in the east. It was the Bravo Shot, the largest detonation ever produced by the United States: an H-bomb equal in output to 1,200 of the bombs dropped on Hiroshima. Eyewitness accounts reveal bewildered awe and terror at this unannounced spectacle (Barker, 2004: 51–53), dramatic proof of the fire-bringer's might (Carucci, 1989). Equally worthy of the trickster image was the long history of nuclear coverups and broken promises. The Rongelapese had received no advance warning of the

Bravo Shot, nor had any islanders been told of the cancer and birth defects that would result in many communities from that and other detonations. Those who were relocated prior to testing were left with the false expectation that their homeland would still be livable when they returned. To islanders it became obvious that the Americans were either cagey or downright duplicitous about the fallout from their testing program: they gathered evidence of radiation without informing locals of the reasons or results (Dibblin, 1988: 52) and continually vexed islanders with inconsistent and inadequate information about the extent, longevity, and effects of radiation (Barker, 2004: 99; Dibblin, 1988: 66; Niedenthal, 2001: 107–108). Many Marshallese now feel that, whatever American officials may say, the entire archipelago, not just the northern atolls closest to ground zero, were contaminated by the blasts, and people attribute all manner of ills to lingering radiation: a 2008 flood, discolored breadfruit, diabetes, short lifespans, hermaphroditic pigs, and declining arrowroot productivity. The US has atoned in the form of nuclear reparations, cleanup costs, and medical treatment, but most islanders feel that the US has done so reluctantly, belatedly, and inadequately—an issue that rages to the present day. Like Ḷetao, the Americans had swindled the islanders by promising to provide, only to bake the chief and disappear.

So in Marshallese discourses of cultural change, it is foreigners who triggered the decline. Marshall Islanders sometimes lay blame on those foreign groups: for the damages of Chinese immigration, Chinese people themselves are seen to be at fault; for nuclear exile and irradiated soil, Americans (or US government officials) are, naturally enough, the culprits. But—and this is a point both crucial and counterintuitive—the prevailing tendency in the Marshall Islands is actually in the opposite direction. When asked who is to blame for cultural decline, locals overwhelmingly[8] blame Marshall Islanders themselves. "Us Marshallese. We cannot blame foreigners." "It's Marshall Islanders' fault for imitating Americans (*kappāllele*)." "It's adults' fault for not teaching their children." "It's *not* America's fault! It's *our* fault. Ourselves. *We* don't eat breadfruit. It's because we're lazy—American food is so easy to cook." "It's Marshall Islanders' fault, because we abandon tradition (*joḷọk ṃanit*)." Foreigners were only the catalyst: culpability is seen to lie on those who too eagerly drank from the poisoned chalice. The same viewpoint is clear in public discourse: the speakers are too busy castigating their fellow Marshall Islanders for adopting American culture to bother condemning Americans for having brought it. It seems, then, that the moral focus of the legend of Ḷetao and the Gilbertese chief is not the trickster's indecent proposal but the chief's and people's acceptance of it.

In this Marshallese discourse of in-group blame, one can see the pragmatic adoption of inward-facing morality by a small nation that can hope only to change itself rather than to influence others; a productive effort to make issues locally meaningful because locally caused; a missionary legacy of Christian guilt (see Carucci, 1984: 146) reapplied to the realm of tradition; or the results of "symbolic violence" (Bourdieu, 1994), the duping of a missionized, imperialized, militarized, marginalized people into the false consciousness of self-accusation. Whatever the origin, this in-group blame is a crucial dimension of the discourse, leading to both its parochial nature and its opportunity for local engagement, meaningfulness, and action.

To summarize in terms of a western myth: *M̧anit* is Eden; westerners were the serpent; money was the forbidden fruit; and the original sin was to be seduced by those foreign things, to eat the apple of modernity. For this fall from grace, blame is focused on the tempted rather than the tempter, just as, in the Garden of Eden myth, the serpent's punishment of being cursed to slither on the ground seems secondary to that inflicted upon Adam and Eve, who are exiled from the garden and condemned to the pains of childbirth and toil.[9]

Any ethnographer worth his salt must be wary of any simple statement of "what the Bongo-bongo believe." While I will focus on "modernity the trickster" in this book, for reasons to be spelled out shortly, it is not the only notion of foreign influence and cultural change held by Marshall Islanders. There is, in fact, a pointedly contrary conception of modernity as progress and America as *savior* and *benevolent chief*, as other Marshall Islands anthropologists have explored (Carucci, 1989, 1997b; Kiste, 1974; Walsh, 2003). These positive sentiments rest on an alternative narrative of Marshallese history, beginning with a rosier view of American arrival. As an outer island man told me, echoing the words of many other Marshall Islanders:

> It is better now. In 1857, the missionaries came. . . . They brought the light. When they showed people the Bible, then there was enlightenment and the fighting stopped. Now people take care of each other. Now they no longer kill each other. They used to kill each other very often.

Here conviviality was created, not destroyed, by foreign contact, and the Americanization of the islands was a welcome development.[10] Similar views are voiced in speaking of the American defeat of the Japanese during World War II. Everyone that I (and others—see Carucci, 1989: 75) have interviewed narrates that arrival in tremendously positive terms: it was akin to the advent of missionaries, for the Americans are credited not only with liberating the people from Japanese cruelty and bringing crates full of delicious food to starving locals but also with having reinstated the Christian services that the Japanese had banned during the war; the gratitude is expressed today at Liberation Day ceremonies on many atolls.

They were Christian saviors—and benevolent chiefs, too. By showing military might (defeating the Japanese, "the most powerful people in the world") and giving food abundantly, America had checked both boxes on the chief's job description. Indeed, the savior image and the chief image are hard to distinguish, since many Marshallese see American power and wealth as a sign of God's favor. Having bested the Japanese in both largesse and strength, America earned the loyalty of the Marshallese as a new "surrogate for the . . . chief" (Kiste, 1974: 88; also see Carucci, 1998), whose authority was not resented as long as it brought sustenance with it (Walsh, 2003: 17–18). Although the United States was at times remiss in its chiefly obligations—in particular by neglecting those displaced or irradiated by nuclear testing—it often, in the local view, did its duty, even after the country achieved independence: it has underwritten the better part of the country's economy, provided disaster relief, ensured the country's defense, and provided humanitarian visitors and Peace Corps volunteers. Even the atomic war games did as much to reinforce the chiefly image as to undermine it: demanding of the people large sacrifices (indeed for some islanders the ultimate sacrifice: their land) in order to secure military advantage from foreign enemies was among the chief's key duties.

Flattering views of America the savior chief are evident in Marshallese society. Marshall Islanders almost always like individual Americans: US visitors are warmly welcomed into Marshallese homes, given countless privileges and positions of honor, and forgiven for cultural peccadilloes. Even anger at nuclear testing is often remarkably muted. Bikinians are notably friendly to Americans, even those to whom they have no incentive to ingratiate themselves. A man on Leb told me that his island had been irradiated by Americans, yet he also told me approvingly that it was God's love of Americans that gave them their preeminence in world affairs, that American scientists had magnificent powers of intelligence and perception, and that Marshall Islanders were glad to live under the protective hand of the United States. A Marshall Islander posting on Yokwe.net opined: "As long as we are attached to US government, we will always be safe and protected." The Bikinian flag is nearly identical to America's, communicating loyalty rather than opposition. Most Marshall Islanders (though by no means all) are unperturbed by the military base on Kwajalein Atoll, and a man on Ujae told me that he hoped America would build a base on his atoll so that his people, too, could benefit from the jobs and rental payments. Both the government and citizens at large, publicly as well as privately, make statements in favor of the US's War on Terror, expressing solidarity in the Afghanistan and Iraq wars. All describe 9/11 as an unprovoked outrage committed upon a good people, never as a deserved retribution for imperial abuses. This support is more than symbolic, as more than a hundred Marshall Islanders serve in the US Army (Journal, 2004b). (This is eerily similar to the traditional pattern in which commoners supplied manpower and chiefs supplied military leadership and know-how.) Citizens refer to these campaigns as a justified fight against evildoers (*rinana*). America's Independence Day is commemorated in the country's newspaper with countless tributes to the United States. While some of this, especially public statements from government officials, could be interpreted as calculated ingratiation, its depth and breadth indicate that it is much more than just that.

In this contrast between modernity the trickster and modernity the savior chief, one hears echoes of the old Pacific double view of foreigners: in the classic case (Obeyesekere, 1992; Sahlins, 1981, 1985), the Hawaiian uncertainty as to whether to worship Captain Cook or kill him. One could simply stop here and conclude that Marshallese attitudes toward America and modernity are ambivalent, contingent, or contested. Be that as it may, it poses a problem for the scholar of risk perception who is looking for a preexisting ideology into which a risk message has been received. If both progress and regress, America the trickster and America the savior chief, are narrated, what exactly is the preexisting ideology by which risks are appropriated according to Douglas's theory? Does it exist at all? One approach to this conundrum is to attribute the divergent views to different factions within the society and then analyze how each faction understands the risk in question. But that will not do in this case: it is usually the same individuals who hold *both* views (Rudiak-Gould, 2010). Another approach is to attempt to reconcile the apparently contradictory views into something consistent, as many anthropologists are wont to do (Brunton, 1980; Gellner, 1970; Rudiak-Gould, 2010). This will not work either, because, as I have argued elsewhere, the views really are contradictory, even to locals (Rudiak-Gould, 2010). Another approach is to throw up one's hands and say that there is no such thing as the Marshallese world view: Marshall Islanders "believe" many contradictory propositions at once, so explanations for their responses are to be found not in the "beliefs" themselves but in the selective, strategic deployment of particular ideas in particular situations (Carucci, 1998; Wynne, 1992: 287); thus risk perception theories based on "prior commitment" are misguided from the start. But this overestimates the chaos in Marshallese ideology, and underestimates the genuineness of people's belief in it: it is not just a rhetorical resource, it is a *conviction*; and while it contains its share of inconsistencies, certain sets of ideas within it *are* consistent, and are wielded as such.

Yet another approach is to take as one's preexisting ideology that set of consistent ideas which are in fact used by locals to interpret the risk one is interested in. Still another approach is to take the *predominant* or *most salient* set of consistent ideas as the preexisting ideology in question and to see how far this can, by itself, illuminate the risk perception under study. In this book, following the methodology of Sheila Jasanoff (2005: 45) in her study of the reception of biotechnology in three western nations, I will adopt these last two approaches. Modernity the trickster is the set of ideas most often used to interpret climate change; this will be demonstrated throughout the remainder of the book. Modernity the trickster is also easily the most discursively dominant and salient set of ideas in the country; this can be demonstrated now. While the narrative of progress (integral to the "America the savior chief" discourse) is as *sacred* as the narrative of regress (integral to the "modernity the trickster" discourse), in that locals refuse to explicitly reject one in favor of the other (Rudiak-Gould, 2010), that does not entail that they are equally *salient*. The narrative of decline is far more so. My informants narrated decline far more often than progress both spontaneously and when asked about the past. In my survey, with 146 respondents, almost 90% considered changes from the past to be a problem, thus implying regress. In the murals described above, artists depict the good old days but never the bad old days, and their portrayals of the Marshallese good life never include depictions of western technology and almost never of churches.

The narrative of modernity the trickster is indeed extremely pervasive, found vigorously in all age groups, among both men and women, educated and uneducated, in outer islands and urban centers.[11] Elderly people may voice this lament a bit more adamantly than young people, but nearly everyone espouses it. In an interview, one need not broach the subject: the informants will usually do so themselves, prompted by associations as diverse as language, immigration, crime, and education. They may willingly discuss this topic for the entire length of the interview, even if other topics are brought up. In a survey result I will present in Chapter 3, changing lifestyles and mores (the narrative of decline that lies at the heart of the "modernity the trickster" discourse) is the second most pressing concern, only narrowly outdone by economic hardship (which is itself often spoken of as an impact or aspect of modernity). The narrative is found not only in captivity, as it were, in the artificial circumstances of an interview or survey, but also in the wild, in speeches on the radio, government documents (RMI Government, 1996), written articles (see for instance Kabua, 2008; Murphy, 2007), letters to the editor in the newspaper (see, for instance, Jormelu, 2007; Nysta, 2003; Rowa, 2001), and many political cartoons. At a conference sponsored by the Customary Law Commission, the main agenda—to reach a consensus on the definition of certain Marshallese customs such as the meaning of a *jowi* (clan)—had to wait until participants had spent the entire first morning of the conference lamenting cultural decline. Such is the power of the narrative. Locals disagree in their level of optimism about how much *m̧anit* can be retained or regained, with a few dismissing it as a lost cause while others believe that a traditionalist revival could accomplish much. They also disagree on some of the fine points of the *content* of this threatened tradition, as I will touch on below. But there is no public disagreement and virtually no private disagreement about the trumping value of tradition, its key attributes, its dissipation, and the need to revive it.

I do not mean to suggest that Marshall Islanders are unthinking, slavish traditionalists. In countless domains of Marshallese life, the lure of *bwiin eppāllele*, "the American smell," "the scent of new things," is strong indeed: for instance, people's desire to have imported foods that are easy to store and cook and have now predominated for decades (see Pollock,

1974), and the wish to have money with which to buy it, or their wish to take advantage of the economic, social, and educational opportunities of the Marshallese urban centers and the United States.[12] There are even cases in which money, supposedly antithetical to Marshallese custom, is openly and proudly presented as a token of conviviality during self-consciously traditional activities (Rudiak-Gould, 2009a: 62–65; also see Bloch & Parry 1989 on ambivalent sentiments toward money). But the traditionalist discourse exists not in spite of but because of this lure: as Ernest Gellner wrote, "moralists, in any field, seldom castigate sins which do not tempt their clientele" (1985: 34). Indeed, it is this tension between the stated desirability of maintaining tradition and the many daily temptations to do otherwise that give the discourse its force, its in-group blame, and its potential to be wielded as a "leverage for change" (Silverman, 1971: 13). To employ the Biblical analogy again: after the fall of man, hope remains. Humans, exiled from the garden, may still regain oneness with God through repentance or a self-disciplined return to righteousness.

Thus the discourse of modernity the trickster has the capacity to inspire action—and inspire it does. It encourages Marshallese immigrants to the United States to retain certain activities seen as key to *ṃanit*, such as *Keemems* (first birthday ceremonies spoken of as epitomizing conviviality) and tributes to chiefs, even though the people no longer depend on or inhabit their land. It is used by educators to argue for the "*ṃajeḷ*ization" of the curriculum (Hilda Heine, personal communication). It is employed by the NGO Women United Together Marshall Islands (WUTMI) to defend the dignity and power of Marshallese women, for instance by arguing that wife beating is unjustified because untraditional, and arguing for the traditional right of women to hold *aḷap* (lineage head) titles. (It is also used by their opponents to argue the opposite, on the same traditionalist grounds [see Journal, 2004a, 2005b].) It moved a local man to write a *Marshall Islands Journal* column (called *Jeḷā kajin jeḷā ṃanit*: "Know language know tradition") that aims to defend the Marshallese language from decline by teaching readers proper usage and ancient words. It inspired the founding of the Alele museum, which houses old Marshallese handicrafts; the establishment of the Historic Preservation Office, which conducts and oversees research into Marshallese culture; *Waan Aelōñin Ṃajeḷ* ("Canoes of the Marshall Islands"), which teaches traditional canoe-building skills to Majuro teenagers partly on the grounds that these skills are in danger of being lost in the urban centers; the maintenance and performance of the *Jebwa* dance of Ujae Atoll, proudly claimed by the people of that atoll to be the country's only remaining traditional dance; and a yearly Manit Day holiday which demonstrates and celebrates traditional seamanship, crafts, cooking, attire, and dance. It is used by those with lucrative claims to land on Kwajalein Atoll, which the United States rents for use as a military base, to shame those who wish to deny them their returns by saying that these opponents have forgotten custom, like so many others. It is used in political argumentation, for instance in the common complaint by the *Aelōñ Kein Ad* ("Our Islands") political coalition that the opposition, the United Democratic Party, has put government above custom in its land-use proposals. This is only a small sample of the ways in which this set of ideas works actively to shape Marshallese society.

Modernity the trickster, then, is not mere xenophobia, grumpy conservatism, or inert nostalgia: it is a sense-making device, a rhetorical resource, an emotional comfort, and a political leverage. Like the Eden myth it both explains the existence of suffering and offers a prescription for its amelioration. It is this cautionary tale that has allowed Marshall Islanders to maintain their distinctiveness against the forces of homogenization—so far.

2 Climate Change Dawns on Marshall Islanders

The apocalyptic future scenarios of climate change in the Marshall Islands have not yet come to pass. By and large life goes on and people do not yet live in constant struggle with its present impacts or in constant terror of its future ravages. Yet climate change marks this archipelago, not only as a burgeoning physical influence on its shorelines and weather patterns but also as an idea, a topic of conversation, debate, fantasy, anxiety, and determination. In June 2009 a group of Marshallese men were passing a morning slowly on Ejit Island, sitting on plastic lawn chairs on the gravel-covered grounds of a house, waiting for the funeral of a local woman to begin. Swapping jokes, offering observations, gently arguing, they had no trouble filling three hours. With the conversation ranging from Bikinian mythology to the politics of nuclear reparations, from the erosion of the Marshallese language to the scientific prowess of the United States, it was only a matter of time before climate change presented itself as a topic of discussion. It was the oldest man of the group who broached the issue: "There's a problem with saltwater intrusion all over the Pacific. It damages taro patches."

A middle-aged man agreed. "Yes. The tide is higher here nowadays," he said. "We get these very high tides most years now. It's what the scientists are saying. The greenhouse effect. The sun is closer, so it's hotter on earth, and the ice melts at the North Pole. This makes the ocean rise—and the Marshall Islands are very low."

A young man disagreed. "But God made a promise to Noah that He wouldn't flood the earth again. The next time God destroys the earth it will be with disease and all sorts of things instead. It will be with a fire from heaven that He destroys the earth, not a flood."

The middle-aged man had a quick response. "It's not God who is doing it." He pointed around the circle of men.

> It's *us*, us people, who are doing it. God gave us people a choice. We decide what to do. And right now we are making bad choices. People are making enormous landfills. Like in San Francisco and Delap [a Majuro neighborhood], they're doing this. When they put all this material in the water, there's less space for the water. So it has to rise. That's

> why it's happening. Everyone in the world is doing this—America and other places too. That's why the world will be flooded. Also, the sun is getting hotter, and ice is melting—that's also causing it.

The young man remained unconvinced. "No. God wouldn't let that happen. If you believe in God, the Marshall Islands will not be harmed."

The middle-aged man laughed at the younger man's naiveté. "God gave us intelligence so that we can do something about it," he argued. "He said, 'Only if you follow me will you live.'"

The young reiterated his point, and then the old man reentered the fray, agreeing with the young man: "Yes, I agree. God wouldn't destroy our country, and make us go to places where we have no money and nothing to live on. The choice he gave us, the one you're talking about, has to do with living forever in heaven. It has nothing to do with a flood."

The debate raged on in the good-natured yet vehement way that characterizes Marshallese conversation in all-male groups. While the issue was not resolved, the terms of disagreement were themselves revealing. The reality or unreality of a climatic threat hinged not merely on whether its effects could be locally seen—climate change as firsthand experience. Nor did it depend entirely on the expertise and wisdom of climatologists—climate change as scientific prophecy. Rather, climate change has dawned on Marshall Islanders through three distinct channels of information: reception of scientific discourses, observation of local impacts, and exegesis of the Bible. This chapter examines how Marshall Islanders have become aware of what we—and now they—call climate change, and how locals triangulate between these different sources of authority in understanding it.

THE FIRST CHANNEL: RECEPTION

There is no saying exactly when the scientific theory of the greenhouse effect first entered the Marshall Islands. It has trickled in for decades. The educated elite, with satellite television, trips to conferences in Honolulu, and emails to relatives living in Oregon, could have been aware of global warming as early as publics anywhere, and even the most remote Marshallese community could hear of greenhouse gases and melting ice caps on the radio. Certainly there was some government discussion of the issue as early as the 1980s (Kluge, 1993: 49). But most Majuro residents report hearing about global climate change only by the year 2000 or afterwards, and the idea has become a significant public preoccupation and grassroots agenda item only since around 2009. In a year spent in the country in 2003–2004, I heard the scientific prediction of sea level rise mentioned just once. By 2007 there was intermittent public discussion, and some informants (both educated and uneducated) were willing and able to broach the issue in interviews. By 2009 interest had grown keen among much of the public and grassroots organizations, though the issue was by no means the most common topic of conversation. During fieldwork conducted in the summer of 2012, this significant though not extreme level of interest had been sustained. By 2012 the scientific concept of climate change was familiar enough to most Marshall Islanders that it could be used as a backdrop to, or a metaphor for, other issues: a cartoon that ran in the April 13, 2012, issue of the *Marshall Islands Journal* shows a woman running from a man while he berates her for breaking her promise never to leave him. An onlooker chimes in, *Kallimur ko rej oktak ainwot an oktak mejatoto* ("promises are changing like the climate is changing").

The sources from which Majuro survey respondents in 2009 said they had heard about the scientific concept of climate change were, in order of frequency of mentions[1]: radio (37% of mentions); word of mouth (15%); television (14%); newspaper (10%); school and university classes (9%); workshops, conferences, meetings, and training sessions (9%); discussions in the Nitijela (Parliament) (5%); and the Internet (2%).[2] The easiest source of information to track precisely is the newspaper. The Marshall Islands has only one newspaper: the *Marshall Islands Journal*, an independent weekly publication widely read by locals. Primarily written in English and also available in Majuro, it reaches the educated urban audience more than others, but with some articles in Marshallese, copies frequently carried to the outer islands, and the information relayed informally by word of mouth, its influence is wide. Many articles on climate change, some in the Marshallese language, have appeared in its pages in the first decade of the twenty-first century. Until 2006, coverage was intermittent, with a small spike in 2001 owing to reporting of the Bonn Climate Change Conference and President Bush's withdrawal of US support for the Kyoto Protocol. The first front-page article on climate change appeared in 2007, after which coverage skyrocketed. By 2009, a total of 44 out of 52 issues of the newspaper had at least one mention of climate change.[3] Through 2010 and 2011 coverage dipped slightly but remained strong. A full tally is presented in Table 2.1.

The *Marshall Islands Journal*'s coverage of climate change has ranged from technical, geopolitical, and foreign (such as examinations of climate policy negotiations overseas) to colloquial, cultural, and local. The newspaper's climate change cartoons translate the science into accessible terms (Figures 2.1 and 2.2). In the cartoon in Figure 2.1, the man is singing, "I love the islands where I was born," the first lyric of the first Marshallese national anthem. In the cartoon in Figure 2.2, the child is crying, "Mommy, everything is so dry. I'm hungry!" and the bird is saying, "The water is so salty." Many journal articles encourage viewers to perceive an upward trend in local erosion (Journal, 2012a), flooding (Journal, 2011a), high tides (Journal, 2010a), heat (Journal, 2011b), and drought (Journal, 2011c), and to link these visible local changes to a global anthropogenic process called "climate change," "global warming," or the Marshallese translation *oktak in mejatoto* (see Chapter 3).

Table 2.1 Climate Change Coverage in the *Marshall Islands Journal*, 2000 through 2011

	2000	2001	2002	2003	2004	2005	2006	2007	2008	2009	2010	2011	Total 2000 through 2011
Front-page article devoted primarily to climate change	0	0	0	0	0	0	0	1	1	2	2	0	6
Other full-length article devoted primarily to climate change	2	8	3	1	4	2	1	10	11	30	25	23	120
Cartoon about climate change	1	0	1	0	0	0	1	0	1	2	2	2	10
Short item primarily devoted to climate change	3	4	5	2	2	1	2	7	5	17	10	14	72
Mention of climate change in an article on another topic	1	3	4	1	9	14	15	29	30	73	52	76	307
Number of issues with at least one mention of climate change	7 out of 52	10 out of 52	13 out of 52	5 out of 52	12 out of 53	10 out of 52	13 out of 52	26 out of 52	30 out of 52	44 out of 52	41 out of 53	38 out of 52	249 out of 626

Figure 2.1 Cartoon from the *Marshall Islands Journal*, February 1, 2002. (Reprinted with permission of the Marshall Islands Journal. [www.marshallislandsjournal.com].)

Figure 2.2 Cartoon from the *Marshall Islands Journal*, February 27, 2009. (Reprinted with permission of the Marshall Islands Journal. [www.marshallislandsjournal.com].)

The journal always encourages concern rather than apathy, and never gives voice to anything that might be called climate change skepticism. If anything it leans toward worst-case scenarios. A 2009 article is entitled *Ene ko retta renaj ibwiji* ("Low islands will be flooded") and quotes a climate scientist as saying that the ocean may rise a meter during the present generation and that relocation from low-lying Pacific islands may become necessary by 2050 (Johnson, 2009). The newspaper has also run articles, usually by non-Marshallese Pacific Islander correspondents, that employ the popular journalistic trope (see Farbotko, 2005; Farbotko & Lazrus, 2012; Mortreux & Barnett, 2009) of presenting countries like Tuvalu and Kiribati as "sinking island nations," already inundated by rising seas, abandoned by the people, or with islanders resigning themselves to inevitable exodus (Bataua, 2009a,b; Journal, 2002; Namakin, 2008). Of course rumors of the demise of these countries have been exaggerated: impacts are not yet overwhelming, life continues, and hardly anyone has left because of climate change damages or fears (Mortreux & Barnett, 2009). But Marshall Islanders have largely been convinced by this journalistic rhetoric: many locals told me that Tuvalu and Kiribati were already being ravaged and evacuated and that therefore the Marshall Islands would be next. One student said, "Tuvalu has already gone underwater, Kiribati will be next, then us."

Some locals have heard about climate change on television, although only a small minority of urban residents have cable television access, there are no local stations, and none of the programs are presented in the Marshallese language. Radio is a more accessible medium: everyone, even in the remotest of outer islands, has radio access, and it is common to tune in zealously for many hours every day. The radio carries locally reported news in the Marshallese language as well as BBC world news in English. Both have broached climate change on numerous occasions. The government's Office of Environmental Planning and Policy Coordination (OEPPC) aired a program on climate change and its effect on water resources as part of Water Day in March 2011.

Marshall Islanders also hear about climate change in various conferences, workshops, and outreach events conducted in Marshallese, in particular since 2009 (Figure 2.3). These events have been organized primarily by a handful of NGOs and government agencies; churches have organized only a few and have been much less proactive in general on the issue of climate change. The events include the United Church of Christ in the Marshall Islands' school of deacons and pastors, which in August 2004 invited participants from around the country to share stories of the local impacts of climate change; the Women's Forum on Climate Change and Clean Water organized by Women United Together Marshall Islands (WUTMI), a Majuro-based NGO, in April 2009; the Marshall Islands Youth for Christ National Youth Convention, which included two presentations on climate change by a representative from the Pacific Council of Churches in June 2009; the Women's Forum on Climate Change and Energy organized by WUTMI in

July 2009; the WUTMI Annual Executive Board Meeting (including three presentations on climate change) in August 2009; and a community festival in the Rita neighborhood of Majuro in October 2011, in collaboration with the OEPPC, focusing on assessing and communicating climate change vulnerability. Jo-Jikum ("Your Place"), a new grassroots organization active since 2010, aims to raise awareness of climate change and environmental issues among Marshallese youth and has organized youth-led, climate change-focused events in Majuro for World Oceans Day and 350.org's Connect the Dots campaign, in addition to organizing a radio-broadcast event in which youth asked senatorial candidates their stances on climate change. When I asked Majuro survey respondents where they had heard about climate change, only 9% of mentions were of such events (and only 9% of respondents said they had attended one), but this surely underestimates the importance of this source of climate change reception, because the information is afterwards relayed by word of mouth, broadcast on the radio, and reported in the newspaper, ending up in those tallies.[4]

Climate change is an increasingly popular subject in the Marshallese classroom, to the extent that one teacher feared that some students had become oversaturated with the message even before reaching high school. A two-day workshop in July 2011 hosted by the University of the South Pacific (USP) sought to train elementary and high school teachers to effectively present climate change to their students. Other events geared toward students included a climate change-themed Education Week organized by the Ministry of Education in February 2009; three summer science camps for high school students focusing on climate change science, adaptation, and

Figure 2.3 Learning about global climate change at the WUTMI Annual Executive Board Meeting, August 2009.

mitigation hosted by USP in 2010, 2011, and 2012; and a climate change-themed summer program for high school students at Assumption School in July 2012. I will describe these events in detail in Chapters 3, 4, and 5.

The climate change message first trickled in, then flooded, with a large surge in local media coverage and educational events in 2009. Importantly, all of the reception sources are telling Marshallese people that they should believe that climate change is real and human-made. Although a minority of locals doubt or dismiss the scientific prediction of climate change, there is no *organized* opposition to its reality, nor do individual skeptics publicly voice their skepticism.

What proportion of Marshall Islanders have heard of this scientific discourse? This is a surprisingly difficult question to answer,[5] though some rough figures can be given. In the survey I conducted in Majuro in 2009, a total of 58% said they had heard of the English phrase "climate change," 45% for "global warming," and 39% for "the greenhouse effect"; 60% had heard of at least one of these three English terms, while 53% could define these terms and 28% could give some explanation of the causation, according to scientific theories. These numbers were probably underestimates in 2009 for the reasons mentioned in the last footnote and certainly are underestimates *now*, with several more years of intensive climate change outreach occurring in the interim. Marshall Islands Conservation Society (MICS) officers, who travel frequently to the outer islands to engage with communities on climate change issues, report that most outer island adults have already heard of the issue. In a survey of high school students at Marshall Islands High School in February 2009, with 94 responses, 96% said they had heard of climate change.[6]

THE ROLE OF WUTMI

In 2009, one agent of climate change communication stood out. WUTMI was a pioneer in educating the Marshallese public about climate change. Launched in 1987, WUTMI is an umbrella organization for twenty-four local women's groups in various communities in the Marshall Islands and two Marshallese communities in the United States and has nearly 1,000 official members. While it focuses its educational efforts on women, its mission is broader. It engages with women as a conduit to Marshallese communities as a whole. Often the group invites influential women from each outer island community to attend a workshop so that they will relay the message to others in their community. Headed by Director Daisy Alik-Momotaro and President Mona Levy-Strauss—both of them well-educated, well-connected, articulate, and visible public figures in Majuro life—WUTMI has assembled a women-led force to complement the mostly male-dominated worlds of the government and the chiefly system.

While this sounds to an outsider like a progressivist movement, WUTMI does not present itself as such. Marshallese women, they argue, were accorded great dignity and status in past Marshallese culture; the abuse and disrespect they now experience are the result not of adherence to a sexist tradition but rather the abandonment of a nonsexist tradition. WUTMI thus presents itself not as revolutionary but as traditionalist, even when promoting apparently "progressive" causes; the only change WUTMI advocates is a move back to what they argue to be the traditional role of women. For instance, there was a controversy (see Journal, 2004a, 2005b), ultimately played out in court, over whether a woman can be deemed *aḷap*, head of her *bwij* (matrilineage) and steward of its land tracts. WUTMI argued that women ought to have such a right on the ground that they were traditionally granted it.[7] Indeed, one of the stated missions of the organization is to protect Marshallese culture, an effort that is spoken of as being able to address all sorts of modern deficiencies in education, health, economy, and environment. Earlier WUTMI programs have dealt with human rights (*maroñ in armej*), violence against women, substance abuse, sea turtle conservation, and childhood nutrition.

In 2009, with ongoing programs in voter education, computer training, sustainable livelihoods, and early childhood parent education, WUTMI decided to tackle climate change. Rather than confronting the issue on its own, they have combined it with other issues; and rather than advocating a sense of disempowerment or resentment at foreign countries, they have focused on what Marshall Islanders themselves can do now to deal with the problem. In April 2009, the Women's Forum addressed climate change along with clean water, emphasizing the intersections between the two. The July 2009 forum combined climate change material with discussion of fossil fuel dependence and unsustainable local development activities. The August 2009 executive board meeting, essentially a workshop, approached climate change through the issue of energy conservation. At the latter two events, I presented in Marshallese, linking climate change action with the defense of *ṃantin ṃajeḷ*.

As I will discuss in the next chapter, these workshops appear to have conveyed the notion of climate change in a persuasive and culturally resonant manner. The very noticeable jump in societal attention to climate change between my fieldwork in 2007 and my fieldwork in 2009 probably owes much to the efforts of WUTMI.

THE ROLE OF MICS

Another outreach organization deserves special mention. The Marshall Islands Conservation Society (MICS) was founded in 2004 by British expatriate Steve Why. Under Why's directorship, the NGO presented itself as an essentially western-style conservation organization dedicated to preserving biodiversity in the Marshall Islands—in a nutshell, protecting nature from humans. The focus has shifted since Albon Ishoda, a Marshall Islander, took over as executive director. Rather than protecting the environment from local communities, MICS seeks to protect local communities from environmental degradation by safeguarding the natural resources upon which the traditional subsistence lifestyle depends. People and culture became central to the organization's mission, and conservation was reimagined as a means to an end.

MICS has never seen itself as *exclusively* focused on climate change. The organization perceives and addresses many threats to subsistence: the overexploitation of fish, sea turtles, coconut crabs, and particular bird species; local degradation of reefs; mismanagement of water resources and solid waste; abandonment of traditional marine protected areas; coastal deforestation; and changing values. But beginning in 2007 and in full swing by 2012, MICS officers have come to regard climate change as the central agenda item—a hub concept with spokes leading to all others. Branding an organization or initiative with the words "climate change" of course has enormous benefits for tapping into the now considerable funds earmarked for such projects by foreign donors. But the organization also has more sincere, less calculated reasons for mainstreaming climate change into its activities, as I will explore in Chapter 5.

In terms of climate change outreach, 2009 was WUTMI's year. Since then, though, MICS has proven itself the most dedicated grassroots promoter of climate change awareness and adaptation. (It rarely interests itself in mitigation.) It is now the organization most directly involved with translating foreign academic and policymaking concepts like *vulnerability, resilience, adaptive capacity*, and *adaptation* to Marshall Islanders. One of MICS's most noteworthy outreach programs, and one of the only campaigns by any Marshall Islands organization to communicate climate change to outer islanders, is its yearly training course. In 2007, MICS Public Awareness Manager Milner Okney began visiting outer island communities, identifying interested local partners (called coordinators) and maintaining contact with them by radio. This has blossomed into a large network of coordinators; by 2012 MICS had partners on fifteen atolls. In 2008, MICS began bringing these partners to Majuro each year for three-week training courses. Under the themes of biodiversity and food security, participants completed training in coastal management, well building, and aquaculture. The 2012 training course explicitly aimed to link all of the previous year's topics with climate change. Okney took the participants on a tour of Majuro's urban center, pointing out areas where saltwater had intruded into the soil and where graveyards and seawalls had eroded; he attributed these damages—along with droughts, ocean acidification, sea level rise, and altered seasonality—to climate change. That scientific concept thus became a unifying explanatory framework for the many puzzling changes outer islanders had observed on their islands. MICS officers also present climate change directly to rural communities during their many trips to the outer islands, often with the aid of a series of posters developed by Micronesia Challenge contributors that vividly illustrate the causes of climate change, its impacts on coastal villages, adaptation measures, and possible negative and positive future scenarios. MICS staff report that most outer islanders have already noticed impacts such as erosion and coral bleaching and in most cases have also heard of climate change. MICS's role, then, is to link the two together, tying concrete visible change to the more ethereal process of global warming.

MICS can therefore be understood as a conduit between the local and the global, a pivot between the foreign and the native. It is also a much needed bridge between local communities and the national government. MICS aims to assist each outer atoll in the development of a Resource Management Plan under the Reimaanlok National Conservation Plan framework, a government-supported multiagency project that MICS now heads. In many of its projects it works closely with government agencies like the Marshall Islands Marine Resources Authority (MIMRA) and the Environmental Protection Authority (EPA) and the cross-agency Coastal Management Advisory Council. Most famously, MICS assisted the community of Namdrik Atoll in drafting a resource management plan and founding a community-based organization to implement that plan, an effort that was rewarded in 2012 with the United Nations Development Programme's Equator Prize.

Kobedia ("Where are you?") is the name of MICS's specifically climate change-related initiative. According to MICS staff, the name is intended to challenge Marshall Islanders to reflect on whether they are acting on climate change or merely spectating; it also calls attention to the value of the Marshallese homeland. By tying climate change to subsistence, that lifestyle which in turn ties people to the land and to tradition, MICS has been successful in translating the science of global warming into locally engageable terms. I will examine this further in Chapter 5.

THE SECOND CHANNEL: OBSERVATION

Marshall Islanders, of course, are not only avid radio listeners, newspaper readers, and workshop attendees; they are also keen watchers of and participants in their local environment (Figure 2.4). Daily life, whether in the urban centers or the outer islands, affords many opportunities to register local change that could be linked to global warming. In this country one is always near the water. In Majuro no house is farther than 250 meters from the sea, and the vast majority of homes are within 100 meters of it. At any time of year people must pay close attention to the sea level; different tides lend themselves to different fishing techniques, to launching and beaching canoes and motorboats, to crossing fringing reefs in boats, and to walking between islets on the exposed reef without being stranded or washed out to sea. Seasonal rhythms must be monitored. During *añōneañ* ("north wind") season (roughly Northern Hemisphere winter), people must be wary of drought and of wave damage, in particular urbanites whose dwellings lie perilously close to the shore. During *rak* ("south") season (Northern Hemisphere summer) outer islanders must prepare preserved breadfruit (*bwiro*) to last them through the *añōneañ* season and any shortages of imported food that result when funds are insufficient or supply ships are late or understocked. Seasonality also predicts breadfruit harvests, a major crop in the outer islands and minor in the urban centers. Men take into account many sorts of environmental conditions in their foraging activities. They fish from the shore, in the reef shallows around islands, or in motorboats and sailing canoes. They organize overnight expeditions to uninhabited islands to gather sprouted coconut seedlings (*iu*), coconut crabs, lobsters, seabirds, giant clams, sea turtles, and other plants and animals that tend to be scarce or nonexistent on and around inhabited islands. Even in Majuro many men fish, and both men and children collect octopus, clams, and shellfish on the reef. Outer islanders must monitor winds before and during sailing expeditions. Both men and women, but especially the latter, pay close attention to water resources, the freshness of well water, and the fullness of rainwater tanks. Women gather plants to be used for local medicine (*wūno*), a still vigorous practice that relies in part on environmental observation.

Figure 2.4 Watching the ocean, Jaluit Atoll (Photo by the author).

Even those locals who have never heard of climate change commonly report that the environment is changing. In my 2009 Majuro survey, 71% answered that the *lañ* (weather, sky, cosmos) had changed, 72% said the same of the *mejatoto* (climate, air, atmosphere, cosmos [see Chapter 3 and Rudiak-Gould, 2012a]), and almost all considered that change to be a problem. The most common specific change in the *mejatoto* or *lañ* that respondents reported was increased temperature, with 62% of Majuro survey respondents spontaneously mentioning it. This was expressed in various ways, from "It is hotter nowadays" to "The sun is stronger" to "It is very sunny nowadays" (which can also refer to a lack of rain). Almost no one reported that the weather was colder.

Many locals report that the sea has risen. In the Majuro survey, 34% of respondents spontaneously mentioned this change. The observation was expressed either as the seas being higher in general, low tides being nearly as high as high tides, or high tides being unusually high. Though almost no one said that seas were lowering overall, a small number, 5%, felt that both high and low tides were more pronounced than they were in the past. Although I did not conduct a systematic survey in the outer islands, interviews on Ujae, Leb, Mejit, and Jaluit show that many outer islanders, too, have noted sea level rise. Reports were mixed, though, on Leb and Mejit, with some locals saying that sea level had risen and others that it had fallen. On Mejit, there is a legendary woman, now resident in a coral head at the north end of the island, who is credited with having lowered the sea and calmed waves.

Many Marshall Islanders point to increased erosion—14% of Majuro survey respondents spontaneously mentioned this change—and the examples given are almost without number. In Majuro Atoll, some beaches have been stripped of sand, and coastal trees are falling in various areas outside of the main urban area (Figure 2.5). Many families have put up seawalls, in almost all cases expressly to protect their houses from floods, waves, and erosion. Many neighborhoods in Majuro, such as Jenrōk, are lined with a virtually uninterrupted fortress of makeshift seawalls cobbled together from concrete, piled rocks, metal drums, and in some cases discarded cars. Most people reported to me in 2009 that they built these seawalls only recently, with the median response being 2001: less than a decade before I asked the question. This indicates that the risk of erosion has increased significantly in the recent past. (Marshall Islanders use the English word *seawall*—there is no equivalent in Marshallese, attesting to the newness of this need.) Laura, a less urbanized community on the opposite end of Majuro Atoll, has experienced erosion along both its ocean and lagoon shores. A man showed me a plot of land that used to be a garage five years before and had now been washed away. The end of Laura (*ṃaan Laura*), a beach and popular recreational area, is often cited as an area of noticeable erosion. Outer islanders also observe erosion in their communities. Easily visible erosion was shown to me on Ujae Island in Ujae Atoll and

Piñḷap Island in Jaluit Atoll; Marshallese say that the "waves eat the land" (*No̧ rej kañ āneo*). On Ujae in 2007, I was told that the lagoon shore had retreated by 10 to 15 feet in one area since about fifteen years before, and a shipwrecked Japanese fishing vessel resting on the shallow reef had recently begun to erode rapidly. Another outer island community, the semiurbanized island of Jabwor in Jaluit Atoll, has many seawalls, indicating erosion troubles. Residents of Jeh and Jabat Islands complain of coastal erosion, flooding of roads, saltwater contamination of wells, and salt spray damage to crops (Kench et al., 2011: 12, 22). I was occasionally told that entire (very small) islands had been washed away by the sea, such as an islet just east of Kalalin in Majuro Atoll and an island in Wotje Atoll; these accounts were not widely known, however, and I was unable to verify them.

For many locals, the most worrisome effect of erosion has not been the destruction of trees, which is not so extensive as to threaten people's livelihoods, but the erosion of graveyards, which often carry enormous symbolic and emotive resonance. On Ujae Island, erosion on the lagoon shore has more than once uncovered and destroyed ancient graves. On Piñḷap Island, an ancient graveyard called the Bōn has been heavily eroded, as I will discuss in detail in Chapter 3. The fact that this graveyard is many hundreds of years old shows that erosion has reached areas that have been safe since

Figure 2.5 Fallen coconut trees, lagoon shore, Laura village, Majuro Atoll, 2009 (Photo by the author).

Figure 2.6 Remains of the Jenrōk graveyard, Majuro Atoll, 2009 (Photo by the author).

prehistory. I heard other reports of eroding graveyards on Anekallimur Island and Laura. A newer graveyard in the Jenrōk neighborhood of downtown Majuro has been all but washed out of existence by waves (Figure 2.6). In my Majuro survey, a vast majority of residents (88%) were familiar with this dramatic example of erosion, though only a bare majority, 53%, felt that the erosion of coastal graveyards was unprecedented.

Some Majuro residents report increased flooding (14% of survey respondents spontaneously mentioned this change), citing the floods and high wave events that have struck the country in recent years. Noteworthy among these was a series of floods, spaced only days apart, that hit the archipelago in December 2008. In Majuro, the waves ruined seawalls, destroyed at least 20 homes (Radio Australia, 2009), damaged 200 more (BBC, 2008), made at least 600 residents temporarily homeless (Journal, 2009a), and caused at least US$1.5 million in damage—significant numbers in this city of only 25,000 people. Streets and yards were littered with fish, rocks, coral, and, most worrisome of all, hundreds of tons of waste (ABC, 2008; Journal, 2008c). Approximately 130 outer islanders on Arno Atoll had to seek shelter outside of their homes (Hezel, 2009a). An islet of Maloelap Atoll was cleaved in two, while Kwajalein Atoll saw an airstrip and a causeway inundated. Saltwater contaminated soil and damaged crops in several outer islands. On December 24, the government declared a state of emergency. The event was something less than a catastrophe, but it nonetheless managed to touch the lives of a large majority of Majuro residents. Of 146 Majuro survey respondents, 71% said they had witnessed the flood at first hand, 74% reported that their friends' or relatives' houses had been damaged, and 32% said that their own houses had been damaged. High tides struck again in January and February 2011, arising from a combination of natural tidal cycles, La Niña conditions, and long-term sea level rise (Radio New Zealand, 2011b). Damage to houses and crops was concentrated on Kili Island. Majuro, Namdrik, Ujae, Lae, and Wotho Atolls were also reportedly affected (Rowa, 2001). In February 2012, the same causes led to floods that washed out portions of a causeway in Kwajalein Atoll, forcing short-term school closures due to transportation difficulties (Radio New Zealand, 2012). But locals are not fully agreed that flood events have worsened. Although almost no one reports that flooding has decreased, only 25% of Majuro survey respondents felt that the December 2008 flood was unprecedented, and that flood seems to have done little if anything to increase people's concern about sea level rise, climate change, or environmental change (Rudiak-Gould, in press).

In the Majuro survey, very many respondents pointed to a raft of other disturbing changes in the sea, especially Majuro's lagoon: fish and coral are noticeably less abundant than they once were, and the lagoon is now polluted (*ettoon*) with garbage and unfit for swimming. A few Majuro survey respondents reported that there are fewer species of fish, and fish that used to be safe to eat are now poisonous. (Residents of each atoll know which species are safe and which are poisonous, and this is presumed to remain stable over time.) Outer islanders' statements about the decline of marine resources are less dramatic, though even there some locals report that fish and coral are declining and fish poisoning is increasing. No one in Majuro or the outer islands reported that marine resources had become more abundant.

Many Marshall Islanders report decreased rainfall and more pronounced and frequent droughts; 23% of Majuro survey respondents spontaneously mentioned this change. This is expressed either as "It rains less nowadays" or "It is very *det* [sunny and therefore rainless] nowadays." El Niño-induced droughts have struck in 1983, 1992, and 1998. Residents remember these events as times of significant hardship, though not of desperation, as the situation was mitigated by help from the Marshallese and American governments. In the 1998 event, many children in Ebeye—the country's other urban center—routinely missed school in order to take the ferry to nearby Kwajalein, site of the American military base, to wait in long lines to fill tanks of water from faucets connected to the US military's never-failing water supply. At least one child was named El Niño after the event. During the 1983 event, the local band Ladrik in Takinal ("The Sunrise Boys") wrote a song with the lyrics (my translation) "How will it rain hard / So that our water tanks are filled? / . . . We're facing a hardship that is dangerous to this life." More recently, a 2012 drought on Wotho Atoll was severe enough to result in the US government bringing in reverse osmosis water-making machines. Only a few islanders reported that rainfall had increased.

Very many Majuro survey respondents (21% of respondents and a full 46% of respondents over the age of 50), as well as a number of outer islanders, report that the seasons have changed. This was almost always described in terms of wind patterns. The "summer" season, *rak*, is expected to be relatively windless, with calm seas, while the "winter" season, *añōneañ*, is expected to be windy, with large waves and choppy seas. Locals said that the neat seasonal categories of the past no longer held: *rak* often brought inappropriately windy weather and choppy seas, while *añōneañ* was strangely calm. Since the Majuro survey was conducted during the *rak* season, many survey respondents emphasized that recent months had been windy, like an *añōneañ* season. A few were concerned that the usual pattern of a rainy *rak* and a dry *añōneañ* was no longer so predictable, and a few others reported that pandanus trees were uncharacteristically producing fruit during the summer season. Some survey respondents reported that wind intensity had changed overall, across the seasons, though there was no consensus as to whether it had increased or decreased. A few locals reported that wind now comes from many different directions, unlike in the past, and a few reckoned that typhoons were more frequent now than in the past.

Some locals report that plants have died owing to the decline in rainfall, the increase in heat, or saltwater intrusion; that breadfruit and coconuts are smaller than they used to be; or that these fall from the tree dead before they are ripe. Many believe that arrowroot grows far less well than it used to or has disappeared entirely. (This is often blamed on lingering radiation from American nuclear testing but now is sometimes also attributed to changes in the weather.) A few Majuro residents complained that with fewer plants, rainstorms produce a huge amount of mud. Many Majuro residents report new diseases, more frequent illness, and more germs in the air and water. When I asked whether the *mejatoto* had changed or stayed the same, people also described changes such as cultural decline, accelerating time, and a recent solar eclipse. While these changes do not fit comfortably within western categories such as "climate" or "environment," for Marshall Islanders they are often in the same general category (the *mejatoto*) as rising sea levels, altered seasonality, and so forth (see Chapter 3).

The overall picture that emerges is quite consistent: hotter temperatures, rising seas, dwindling marine life, eroding shores, sunnier weather, declining rainfall, perturbed seasonal wind patterns, and the beginnings of odd and adverse alterations in crops. The descriptions are consistent not only with each other but also with scientific measurements. Temperatures are measured to have increased by 0.12°C per decade on Majuro Atoll since 1956 and 0.20°C per decade on Kwajalein Atoll since 1960 (Australian Government, 2011: 4). The ocean has risen in the Marshall Islands by an average of 7 millimeters per year since 1993 (Australian Government, 2011: 4). Rainfall in the Marshall Islands has been decreasing since 1950 (Australian Government, 2011: 4), with more frequent droughts in the dry (*añōneañ*) season. There is also some scientific substantiation of increased erosion in Majuro Atoll (Xue, 2001), though the evidence here is mixed, with accretion as well as erosion in evidence, and local land reclamation and development obscuring the signal of sea level rise-induced erosion (Ford, 2012).

In the Majuro survey, older informants were considerably more likely to report environmental change, even while holding education and scientific awareness constant. This was especially pronounced in the case of altered seasonality, reported by 3% of young people (18 to 29 years), 12% of middle-aged people (30 to 49 years), and a full 46% of elders (above 50 years). This suggests that it is indeed a lifetime of environmental observation that inspires locals to report environmental change; people do not report changes simply because they believe in general decline, nor are they merely parroting media reports of rising seas and hotter temperatures. At the same time, Marshallese reports of change are not hermetically sealed, and there is strong evidence to suggest that the dissemination of global warming discourse has encouraged locals to notice more change than they otherwise would have: quantitative analysis indicates that Marshall Islanders who are more aware of climate science are also more likely to report change, with the direction of causation most likely running from the former to the latter (that is, being aware of scientific discourses of climate change makes people more likely to report local environmental change).

It is difficult to say whether these observed changes are due to global anthropogenic climate change: attributing any single event to global warming is always problematic. The December 2008 flood was caused at least in part by a low pressure system north of Majuro combined with the high tides expected at that time of year (Johnson, 2008); long-term sea level rise was only one factor. As for the erosion in Majuro Atoll, this probably owes less to global warming than to various local human activities (Juumemmej, 2006: 97; Xue, 2001). The US military joined several islands together to make what is now the D-U-D (downtown) area of Majuro: natural channels were thus sealed with landfill, altering currents and exacerbating erosion. The same has resulted from the human removal of salt-tolerant coastal trees, which bind the soil together and protect less salt-tolerant trees from ocean spray. Seawalls may only worsen the problem they are intended to solve: they prevent erosion where they are constructed but increase erosion around them, and they prevent the wash-up of debris that naturally builds up the islands and protects against waves. Extensive dredging has occurred in Majuro Atoll in order to create landfill, and this increases erosion as well: building more land in one place means stealing it from somewhere else (UNFPA, 2009). Thus, some of the environmental impacts I have described likely have causes other than global warming. But other changes—such as sea level rise, increased temperatures, and decreased rainfall—are certainly not the result of local actions, and here global climate change is a likely culprit. In truth, the causation is irrelevant. The important point is that these changes are occurring at all: they are visible local damages consistent with the scientific discourse of global climate change and thus a potential source of information about the threat of climate change.

THE THIRD CHANNEL: EXEGESIS

Marshall Islanders' third font of climate change information is a post hoc source, consulted for guidance once one has already become aware of climate change through reception or observation. This third channel is the Bible. The scriptural interpretations of climate change that I will describe below are not, to my knowledge, officially encouraged by any church or other group. No one told me that they espoused a particular biblical view of climate change because their minister had advocated it. Nor did I ever personally hear a sermon that concerned climate change, despite attending many of them, although reportedly the subject is occasionally mentioned. By all indications, then, Christian interpretations of climate change are improvised individually and circulated informally. The country's largely Protestant heritage makes this possible. Accordingly, I will discuss religious interpretations through the course of interviews rather than by describing particular sermons, ministers, churches, or religious movements.

The Bible is an extraordinarily flexible text; it can be made to speak even to the unprecedented phenomenon that is global anthropogenic climate change. Two books in particular, Genesis and Revelation, lend themselves to such exegesis. In Genesis we find the myth of the Flood, and its applicability to climate change in a low-lying island nation is not hard to fathom. Some Marshall Islanders liken future sea level rise to the Flood, naturally enough given that climate change has often been communicated to Marshall Islanders in a way that equates it, purely and simply, to nationwide inundation. (Recall the newspaper article entitled "Low islands will be flooded" [*Ene ko retta renaj ibwiji*]). The Marshallese term for a very high tide, for a flood, and for the Biblical Flood is the same (*ibwijleplep*) in each case, making the comparison locally intuitive.

What can be done with this comparison? The Flood myth does not end with the banishment of the waters. As the ark settles again on dry land, God establishes a covenant with Noah and all humanity after him: "Never again will all life be cut off by the waters of a flood; never again will there be a flood to destroy the earth. . . . I have set my rainbow in the clouds, and it will be the sign of the covenant between me and the earth." (Genesis 9:11–13) The rainbow, appearing at storm's end to prove that the deluge has passed, continually renews that promise. Some Marshall Islanders cite this passage to question scientists' predictions of sea level rise. For instance, a middle-aged Ujae man of limited education but considerable local savvy told me, "It's untrue that Ujae will be covered with water. . . . Because in the Bible . . . it says [God] won't destroy the world with water, because he already did it at the beginning. . . . I believe God."[8]

Those who invoke the Genesis promise often add that the next time God destroys the earth, he will do so with fire, not water. The reference here is to the opposite end of the Bible. Revelation 16:8–9 reads: "The sun was given power to scorch people with fire. They were seared by the intense heat and they cursed the name of God, who had control over these plagues, but they refused to repent and glorify him." If climate change equals sea level rise, then the scientific predictions have little in common with the prophecies of Revelation; climate change is refuted.

As an alternative or addendum to the Genesis argument, some Marshall Islanders argue that God would never allow global warming (or anything else) to destroy the archipelago that he created as the rightful and eternal homeland of the Marshallese. A middle-aged man in Majuro lived directly adjacent to the ocean shore and had recently built a seawall to protect his house from the waves. But he told me that the waves were no worse than in the past and would never grow larger because Marshall Islanders pray so zealously to God. A well-educated Majuro man told me, when I asked about climate change and God's promise:

> I don't think God wants to punish me, or any other people. [So] maybe I should believe He won't flood the earth. I know there is a higher power. I leave my well-being up to him. For instance, in 2001, I was diagnosed with diabetes. But I don't take medicine for it or ever go to the doctor, because I let God take care of me. And I've been fine—six years of having diabetes, and no problems.

By 2012, Biblical arguments against the reality of climate change appeared to have declined in popularity among Marshall Islanders, but the idea persists among a minority, and some climate change communicators (in particular the MICS officers who communicate climate change to outer islanders) complain that it is, more than occasionally, an impediment to their work.[9]

Other Marshall Islanders reject the Genesis argument on scriptural grounds. God promised never again to flood *the entire world*, but he made no promises about low-lying islands. A well-educated Majuro man told me, when the topic of God's promise came up,

> What many people don't realize [is], while it's true that most coral atolls will be the most severely affected, it will also affect the United States as well—Florida and California. The weather patterns will change. Plains will no longer be healthy—wheat fields, and the local trees in the Netherlands. Those places will be flooded.

It is usually the well educated who make use of this counter-counterargument for climate change, since it requires some appreciation of the global implications of the threat.

Another rebuttal to the Genesis argument goes like this: God's promise was made in the time of Noah, and is part of the Old Testament (*Kallim̦ur eo M̦okta*, "the promise before"). But the New Testament (*Kallim̦ur eo Ekāāl*, "the new promise") supersedes the Old; the old promise is obsolete. Two men argued about climate change in 2009:

X: According to the news, they say the islands will be flooded. You won't be able to cook your food anymore, because the water will be on the land. The islands will disappear. The water will rise up and up.

Y: I'm a Christian. God promised Noah this wouldn't happen.

X: Yes, God said he wouldn't destroy the earth again with a flood. Yes, this is true. But, that was just in one era. And whose era was that? That was the era of Moses and Adam and those people.

Another argument runs: God only promised that *he* would not flood the earth again.[10] He has not ruled out the possibility that *people* might do so. In 2007 I asked Enja Enos, a prominent minister in Majuro, about religious interpretations of climate change. When I told him that some people I had talked to quoted Genesis to dismiss climate change, Reverend Enos laughed heartily and said,

> This is from naïve people . . . I don't think we have to look at the Bible that way. You see, man is the one who is causing this thing. God did not destroy our reef. We are destroying the reef. . . . Like [for instance] making this bridge, the reef has to be dynamited. . . . Things are really changing, and it's not because God said so. . . .

> These development[s] are destroying things. . . . And even the islands that we used for the [nuclear] testing. So many things men have made. . . . *We* are doing it.[11]

An officer at the EPA told me that she often argues with pastors who deny climate change on Genesis grounds. Her rebuttal was that after the expulsion from Eden, humanity has been given the choice to do good or evil, to abuse the environment or to cherish it.

There is yet another rebuttal of the Genesis argument. Yes, God will destroy the earth the second time with fire, not water; but that fire is global warming. Here is a Biblical argument *for* the reality of climate change. Responding to the Genesis argument, a Majuro man told me, "Well, He was talking about a flood. But He said He would bring brimstone and fire from heaven. So what is this warming of the air?" He laughed, letting the implication of the last sentence hang. Milner Okney, of MICS, told me that he uses this argument during outreach activities on the outer islands when unconvinced locals appeal to the Genesis argument.[12] Quite a few Marshall Islanders say that scientists' "climate change" will be part of the End Times or that present climatic changes are a sign that End Times are imminent.[13] An elderly Majuro woman, when asked why the climate was changing, answered, "We think it's like the Bible says. Brothers will fight each other. These changes are happening. We're selfish now, like Americans." Some locals offer sea level rise predictions so extreme—"fifty feet in fifty years," "to the top of the palm trees"—that one wonders if they have Armageddon in mind.

Climate change is hardly the first apocalyptic prophecy to circulate in Marshallese society. Although talk of the end of the world does not reach the fever pitch of Melanesia's millenarian movements (see, for instance, Stewart & Strathern, 1997, 1998; Whitehouse, 1995), speculation does circulate and amounts to more than idle talk. Marshall Islanders who witnessed the unannounced detonation of an H-bomb in 1954 wondered if they had just seen the beginning of End Times or World War III and went to church to pray. In the late 1990s some Marshallese church congregations feared that apocalypse would come in the year 2000. Combining nuclear and Y2K fears, a Marshallese man wrote a piece for the *Marshall Islands Journal* informing his fellow citizens that the book *The Bible Code* had predicted that nuclear war would break out during the year 2000 (Heine, 2000). Another Marshallese commentator wrote a letter to the *Marshall Islands Journal* with a nuclear text and perhaps climate change subtext: global nuclear testing had ruptured the sea floor and disturbed the earth's spin; devastating tidal waves and earthquakes would soon result, destroying all life on earth. These stories represent the zenith of Marshallese narratives of decline, and they help to explain why Marshall Islanders would take seriously the bizarre, preposterous, and true idea that their entire nation may sink in fifty years.

WHO CAN BE TRUSTED?

Reception, observation, exegesis: each source of information makes claims about climate change. But which of the sources can be trusted? The question is an important one for Marshall Islanders as they assess the idea of climate change, since the three sources sometimes make incompatible claims. In the case of exegesis, the answer is simple: God is perfectly trustworthy, the Bible perfectly true. Almost no one in the Marshall Islands disputes this. The puzzle in the case of exegesis is not whether the source can be trusted but what exactly that source is claiming. If God came down from heaven and declared that global warming is real, Marshall Islanders would not doubt its reality. But he has not done so; instead he has given humanity the Bible, which contains only scattered, cryptic, and disputable clues as to the reality of climate change.

In the case of reception, the situation is precisely the opposite. It is clear what scientists are saying about climate change but unclear whether they can be trusted. Climate change is usually presented to Marshall Islanders as "what scientists say," and the existence of a few dissenting scientists is left unmentioned.[14] But, this scientific consensus notwithstanding, it is not clear to Marshall Islanders whether scientists are telling the truth about climate change or even whether they know the truth.[15]

As science and technology studies scholars have argued (Jasanoff & Wynne, 1998; Wynne, 1992), the important question here is how Marshall Islanders perceive scientists as a social group, not just "science" in the abstract. Local views of scientists arise not merely from airy philosophical debates about indigenous versus western epistemologies but from concrete historical experiences with scientists in the flesh. Marshallese encounters with scientists could fill their own book, so I will content myself here with a few interactions that seem to me particularly influential or telling. Scientists—and other powerful outsiders with whom scientists are seen to be nearly synonymous—have repeatedly displayed prodigious power and cleverness, but also moral ambiguity, arrogance, and at times outright treachery.

It began with the earliest extended encounter between islanders and Europeans. During Otto von Kotzebue's first visit to the Marshall Islands, in 1817, the captain feared that one of his crew had been kidnapped by the islanders and ordered a "skyrocket" as a warning:

> Scarcely had the cannon been fired, when a dreadful howling arose in the whole island, which lasted for above a quarter of an hour. . . . All fell on the ground as if struck by lightning when the shot was fired. . . . The fright of last night still operated to-day, so that nobody ventured on board till some of our gentlemen had gone on shore. They had asked a great many questions, what the report and flash were? And when they were told that I had on that occasion paid a visit to heaven, my

> consideration was doubled among them, and they behaved with great propriety. (Kotzebue, 1821: 115)

Islanders reacted with the same fearful amazement to a gunshot (Kotzebue, 1821: 76). Kotzebue's accounts probably exaggerate the islanders' awe, as it flatters western prowess and justifies the colonial project. But even if the islanders were only half as astonished as Kotzebue makes them out to be, their astonishment was still great (Walsh, 2003: 144–145).

What Kotzebue's cannon was in 1817 America's H-bomb was in 1954. In Chapter 1 I alluded to the American scientific coverups, misinformation, and broken promises that Marshall Islanders have endured since the era of atomic testing began in their territory. The result was ambiguous. Either the scientists were fools who could not see radiation effects for what they were or they were geniuses who deliberately withheld the truth (Dibblin, 1988: 42–44, 52, 57, 66; Niedenthal, 2001: 92); either way they could not be trusted. Yet at the same time, scientists had proven their cognitive supremacy like never before in a display of raw technoscientific power that left the earth literally shaking. In the words of one man from Likiep Atoll, recalling the Bravo Shot forty years later:

> We woke up in great terror, not understanding what was happening. Many thought they were witnessing the end of the world. A short while after the sky lit up an indescribable sound shook the island. It was so great that it seemed as if the island would split into a million pieces. Standing outside, the intensity of the sound almost knocked me over. Some concluded that maybe the "Big War" that foreigners kept telling us would happen between the United States and Russia had finally begun. . . . After what seemed like ages, the noise subsided and then we heard a different kind of noise, a whooshing sound that seemed to travel at tree-top level. People rolled around on the ground crying and clutching their children to their chests. Even the dogs and other animals were howling and yelling. Then suddenly, there was total silence—no sound except the sound of people crying, dogs barking, and waves breaking on the ocean side of the island. We picked ourselves up from the ground, amazed we were still alive. (Barker, 2004: 52–53)

Suspicion and resentment of scientists mingled with feelings of awe and dependence, as Marshallese often felt they lacked the expertise to measure radiation and predict its effects, relying instead, by necessity, on the mysterious instruments and arcane knowledge of radiation scientists. A Bikinian man expressed it as follows:

> I apologize to [my family] because I don't quite understand the depth of the situation here on Bikini. I am an uneducated man. I am Marshallese and I can't quite understand or tell what is safe and what is

> unsafe here. I can only have faith in the U.S. government. They have the responsibility for telling us what is good for us and what is dangerous. But for myself, my foresight and my knowledge concerning these radiation issues ends right here in front of my face. As I said this morning to those newsmen: "I can't tell if these Americans who are working on this island are doing a poor job, or performing miracles of science. I am uneducated in these matters. I am unintelligent because I didn't go to school to study radiation science. So, I can only hope that the U.S. government will tell us the truth about Bikini, whether it is safe for us to live here now or in the future." (Niedenthal, 2001: 111; also see Barker, 2004: 103)

Nuclear testing thus forced Marshall Islanders into an awkward state of cognitive dissonance: scientists were both the inflictors of pain and the means to relieve it, the bringers of unpredictability and the only source of certitude.

The 1969 moon landing had a similar effect. Almost half a century later, Marshall Islanders continue to cite the event as the example par excellence of the astonishing might and brainpower of scientists and Americans. It is often said, "Americans are smart—they flew to the moon." It was a stunning proof of power, a Godlike act; yet also proof of treachery and a high affront to God. As an elderly man from Ujae told me, "Scientists become arrogant. They want to have Godlike knowledge. They helped people to get to the moon. But God didn't put people on the moon. He put them on the earth. They aren't supposed to be on the moon. They want to know more than God."

Forty years almost to the day after the moon landing, American scientific power played out in the cosmos once again. In July 2009, the *Marshall Islands Journal* and Marshallese radio started carrying stories of a strange scientific pronouncement: on the twenty-second of that month, said scientists, at a precise time for each affected island, the sun would turn dark for several minutes. Almost everyone in the Marshall Islands heard this prediction of *marok eo* ("the darkness"), and the event was eagerly anticipated. When the full solar eclipse came as promised, people were impressed not just by the phenomenon itself (see Chapter 3), but by scientists' uncannily accurate prediction of it. Afterwards, Marshall Islanders frequently cited that prediction as proof of scientists' expertise and trustworthiness, and often added that the idea of climate change must therefore be correct. But outside the path of totality, not all of the onlookers were so impressed. There the eclipse had been only partial: the sun looked as it always does and the world turned only slightly darker. The radio and newspaper had failed to clearly inform people that the path of totality would pass through only some of the atolls, and that islands outside of this path would experience only modest darkening. Also, many people had thought that the scientists had predicted truly nightlike darkness that would last until the next morning. On Arno Island, Arno Atoll, a number of families had stayed in their homes during the eclipse: they ordered the children indoors, closed the doors and windows, and announced that the day's work was over, night was coming, and it was time to sleep. So when the darkening turned out to be slight and short-lived, some took this as a disconfirmation of the scientific prophecy (Elise Berman, personal communication). Some islanders had insisted before the eclipse that they did not trust the prediction: they would believe it only if God (not scientists) told them it would happen. When the partial eclipse occurred, they felt that their suspicion had been confirmed. Other Marshall Islanders outside the path of totality, such as people I talked to on Mejit Island, were nonetheless impressed by the event and took it as an example of a successful prediction by scientists.

Once again the result had been ambivalence. The eclipse probably strengthened trust in scientific expertise more than it weakened it, but both trust and skepticism were outcomes. The event may even have changed the geography of scientific trust along the path of totality, with communities within that path (Leb and Kili Islands; Jaluit, Namdrik, Ujae, Lae, and Eniwetok Atolls; and most of Ailinglaplap and Namu Atolls) convinced that scientists had triumphed, while communities outside of that path were less convinced or even convinced of the opposite.

What understandings of scientists as a social group resulted from these and other historical encounters? Scientists, to begin with, are considered to be *foreigners*. Some Marshall Islanders have scientific qualifications, but when locals talk about scientists, they are almost always referring to non-Marshallese. The Marshallese language has a word for a weather forecaster (*rikatu*), but this is not used to refer to the foreigners who talk about climate change and the coming flood: they are *jaintiij*. A *rikatu* must be Marshallese, while a *jaintiij* is presumed foreign. Marshall Islanders sometimes say that scientists tend to be or should be bald, because they think so much that their hair falls out. This reflects and reinforces the idea that scientists are foreigners, since baldness is rare among Marshallese. More specifically, scientists are assumed to be American (*ripālle*) unless otherwise stated. The most common Marshallese word for *scientist* is in fact a loan word from English: *jaintiij*. Of course, the two categories, *jaintiij* and *ripālle*, are not identical: Marshallese recognize that not all Americans are scientists, and vice versa. But in informal reasoning the two categories verge on being interchangeable. Frequently my questions about scientists would be answered with statements about Americans instead. The two groups are never contrasted with each other ("Americans are like so, while scientists are like so"), whereas, for instance, Marshall Islanders are frequently contrasted from Americans, and Christians are sometimes contrasted from Americans. None of this is surprising because, as we have seen, it was foreigners, usually Americans, with whom Marshall Islanders had their formative scientific encounters.[16]

So Marshallese attitudes toward scientists reflect their attitudes toward Americans. As I explored in Chapter 1, while Marshall Islanders are extremely hospitable toward individual Americans, much Marshallese discourse imagines *ripālle*s in the abstract as powerful and clever but also morally problematic and devious: they give something other than that which they promise. This gives Marshall Islanders reasons for both credulity and skepticism: power and cleverness suggest that scientists know of which they speak, but moral problematicity and deviousness suggest that scientists may not be forthcoming with their knowledge and may prove dangerously overconfident in their predictions.

The former view comes out in a Marshallese image of scientists that I call *scientists as magicians*. Scientists and Americans are credited with extraordinary, possibly unlimited powers. People frequently refer to the "tools" (*kein jerbal*) or "proof, evidence" (*kein kaṃool*) of scientists: vaguely described, mysterious objects of power, revealers of hidden knowledge. Scientists are often called "smart people" (*rimālōtlōt*) and this cleverness is contrasted to a Marshallese lack of understanding and technological expertise. When I asked an elderly man in Majuro what people could do about climate change, he said that perhaps *scientists* could fix it. When I asked him if Marshall Islanders could fix it, he replied, "No, they can't. We don't understand climate change. We have no knowledge. We have no tools." A woman in Majuro asked me if scientists might be able to somehow raise the reef in order to beat sea level rise, and a woman on Ailinglaplap told me that the process by which a local swamp had filled in with land must have been an act of God, because Marshall Islanders could never make an island—*but perhaps Americans could.* The belief in American and scientific omnipotence also shows up in the incredible range of powers that radiation (as a product of atomic bombs built by American scientists) is credited with and in the idea that the December 2008 flood was caused by American military activity.

Closely related is a view of *scientists as soothsayers* who can predict future events with startling accuracy. The 2009 eclipse is now a quintessential example of this, but people also cited past typhoons that scientists had successfully prognosticated. Another flattering view of scientists is *scientists as God's chosen ones*: the notion that scientific power is so prodigious that it must surely be God-given (see Carucci, 1997b: 148). Some Marshallese say, "God gave intelligence [*mālōtlōt* or *kōḷmānḷọkjeṇ*] to scientists—that is why they are smart." A young, vocally Christian man on Leb Island told me, "Americans defeated all other countries in battle, because they pray a lot to God. Before a war they pray to God and ask that He will let them win." In this Marshallese view, military success is a sign of godliness. The man continued, seguing easily from the category of Americans to scientists (*jaintiij*): "The tools of *jaintiij* are very good. *Jaintiij* are very intelligent. They can see all sorts of things. They can see small things with microscopes. They can see if things are irradiated or not." A woman in Majuro told me, "God gave intelligence [*kōḷmānḷọkjeṇ*] to scientists, so we trust them. For instance, doctors, so that they can help people who are sick. All things are from God—He made the world." While talking to a woman on Mejit, I attempted to force a choice between God and scientists as sources of authority by asking, "If God says one thing and scientists say another, who is correct?" But my interviewee subverted the question, answering that in that case "God *and* scientists are correct, because who gave intelligence to scientists? God did!" In this view, American/scientific power is evidence of God's favor.

But that same power can be understood in very different terms, as an affront to God. Scientists, with all their world-shaking power, are the ultimate sinners, the quintessential non-Christians.[17] This is the view I call *scientists as God's rivals*. A man on Ejit reasoned that the Marshall Islands was probably safe from climate change because God would not destroy a Christian country, but he might destroy America with climate change because Americans are arrogant (*eutiej būruer*) and do not follow God. Another frequently voiced belief was that scientists ought not to predict the future; knowing the future is God's work.

A related view is *scientists as tricksters*. Americans are compared to the legendary trickster figure Ḷetao, so scientists, too, can take on a trickster image. At the end of Ḷetao tales, the storyteller will often say, "And then Ḷetao sailed to America and lived there—and that's why Americans are so smart." (Some storytellers add, "but lie so much.") The Marshall Islands ethnographer Laurence Carucci reports that the people of Eniwetok Atoll tell this origin story of American power in more detail. After Ḷetao's wily escapades in the Marshall and Gilbert Islands, he stows away on an American ship. The Americans capture him, but Ḷetao bargains for his freedom, offering his captors the secret to his powers. The Americans take the deal, and that is how they have been able to develop advanced technology such as nuclear bombs, satellites, and airplanes (Carucci, 1997b: 148–149). In this view, American/scientists, like Ḷetao, are as subversive as they are clever. *Scientists as tricksters* appears in attitudes toward radiation—the widely held (and entirely justified) suspicion that American scientists have attempted to fool Marshallese into believing their islands are safe. An elderly man from Ujae was of the belief that arrowroot had begun to grow poorly on his island because of nuclear fallout; he also cast doubt on scientific explanations of diabetes:

> Around 1959, diabetes started to appear, whereas people before 1959 didn't have diabetes. The doctor says, "Oh, maybe there was a diabetic in your family." Sometimes they say, "You eat too much rice. You eat too much meat and fish and use too much sugar." But, I'm not so sure, because in America [where people eat these same things], there are lots of people that don't have diabetes. Many Marshallese people say, "The [real] reason is that we're irradiated." Radiation from the bombs.

Another critical view is *scientists as overconfident braggarts*. Overtly Christian Marshall Islanders sometimes refer mockingly to the scientific claim that "humans came from monkeys"; if they believe such a ludicrous and unbiblical thing, surely they cannot be trusted. On the Marshallese website Rimajol.com in 2007, a Marshallese man posted a comment: "I don't believe we came from monkeys and I don't believe it when scientists say that people are the ones who are making changes in the atmosphere and *mejatoto* [climate, air, atmosphere, cosmos]." Some Marshall Islanders also feel that scientists greatly overestimate their powers of prognostication. Scientists cannot predict the future, even in principle, because doing so is the domain of God, not humans. An elderly woman in Majuro told me, "They say, 'I know about everything.' But I say, 'No—God's plan!'" She said she didn't believe in climate change because "I only believe in God's plan." Other Marshall Islanders say that humans may in principle be able to predict future events, but scientists have tried in the past and failed. Scientists had warned of a Y2K calamity that never came to pass. The Majuro weather station had predicted a rainless month during the rainy season of 2009, and a huge rainstorm had promptly struck. In 2008, inhabitants of Wotje Atoll doubted the reality of climate change because scientists had arrived in the late 1970s and told them that in thirty years, sea level rise would swamp their islands; when this failed to occur, they lost confidence in the predictions (Jacob Appelbaum, personal communication). A Majuro woman told me the following in 2009, when I asked about scientists' idea of climate change:

> They said in 1980 that there would be a drought. But then it rained! God controls everything. I don't believe in climate change because only God controls the *mejatoto* [climate, air, atmosphere, cosmos]. They said there would be El Niño and drought—the EPA said to conserve water for that reason. But only God is in control. Then they said that it would be really dark [the solar eclipse]—so we got flashlights to prepare. But then it wasn't very dark at all—it was so light out that you could walk around and see people! It was just a little dark, like when it rains. Now they say there will be a flood, but God says no—there will be no flood, and the rainbow is proof of that. There will be only brimstone coming from the sky—no flood! Only God can speak about his own works.

Scientists' mixed record of prediction; the wonders they create and the horrors they inflict, often at the same time; their dual role as awe-inspirers and tricksters; and Marshallese puzzles about whether Americans are to be trusted or feared, whether Americans are the quintessential Christians or the world's biggest sinners, all converge toward a complex stew of attitudes toward scientists. What is clear, though, is that Marshall Islanders cannot be confident in the prophecy of climate change simply because scientists have asserted it. There is no perfect certainty to be found here.

Finally, what of the trustworthiness of the third source of information, observation? Marshallese elders in particular are often quite confident in their statements that seas have risen, the air is warmer, rainfall has lessened, and so forth. But the observation that sea level has risen since one's youth is radically different from the prediction that the ocean will continue to rise and eventually inundate the entire country. It is the *future* on which the idea of climate change is most intensely focused (Swim et al., 2009: 36; Taddei, 2009a), particularly in a country whose eventual habitability is at stake, and Marshallese observations of environmental change do not speak, or claim to speak, to that dimension. Marshall Islanders in fact often doubt that they can, even in principle, predict the distant future. God certainly can; "Only God knows what will happen," it is said. Scientists perhaps can, at least according to the *scientists as soothsayers* image. But locals probably cannot, at least where the distant future is concerned. *Katu*, Marshallese weather forecasting, was traditionally used to forecast weather in the short term, for the purposes of planning or aborting open-ocean canoe voyages (Haddon & Hornell, 1975: 374); it has never been intended to predict meteorological conditions for fifty or a hundred years in the future, as climatologists purport to do. Neither has *bubu*, Marshallese divination, which in any case is more often used to ascertain what is auspicious or what is already true rather than to predict the future per se. Those who challenged scientists' prediction of the July 2009 eclipse did so on the grounds that only God could predict such things; it was *divine* foreknowledge, not Marshallese foreknowledge, that was considered capable of refuting scientific predictions.

So vis-à-vis the truth or falsity of climate change, there is no certainty to be found in observation either. Locals find themselves in an epistemological predicament: no one perfectly trustworthy (God) is taking a clear stance on the issue, and those who are taking a clear stance on the issue (scientists) are not perfectly trustworthy. An allegory sums up the situation well. You are standing in a desert, barren and forbidding as far as the eye sees. On the other side of a deep gully is a savanna with fish-filled rivers and fruiting trees. You desperately wish to cross, but the only way over the gully is an old and frayed rope bridge. Near this bridge stand three men, the only people for miles. You ask them if the bridge is safe to cross. The first man to answer is a foreigner of prodigious knowledge and skill but also a well-known braggart with a spotty reputation and uncertain morals. He loudly proclaims, "No! Absolutely not," and discourses at length on the bridge's danger, using strange words and concepts that you only partially follow. The second person to answer is a humble old man who has lived near the bridge since his youth. He quietly says, "I've noticed the bridge becoming more threadbare since I was young, but it might be safe. I can't be sure." The third man to answer is a fellow of perfect integrity, utmost honesty, and limitless knowledge; he has never been known to err or deceive. When asked, he mumbles, almost inaudibly, "Yes. No. Perhaps," and then leaves, refusing to say anything more on the matter. Should you cross the bridge?

The barren landscape is, of course, the prospect of devastating climate change; the abundant landscape is the possibility that global warming is untrue and that Marshall Islanders will be able to enjoy their country long into the future. Marshall Islanders would love to cross the bridge. As we will see, a minority manage to by dismissing the braggart's warning, discounting the humble man's whisperings, and heeding the wise man's first statement. But others are unwilling to cross, as more or less credible warnings are coming from all three parties. In the next chapter I will attempt to explain why, despite this multidimensional uncertainty, most Marshall Islanders do accept that climate change is real and worthy of concern. For now, it is the uncertainty itself that I would like to dwell on, the triangulation that Marshall Islanders perform between three sources of climate change information. This is how Marshall Islanders become aware of what we, and now they, call climate change. The interview excerpts that follow will give some flavor of this tangled process of realization.

BECOMING AWARE OF CLIMATE CHANGE

Fishing from One's Living Room

Middle-aged, well-educated man, Majuro, 2009:

PRG: Will Marshallese people stay in the Marshall Islands in the future?

X: Well, according to those learnings, in fifty or a hundred years the Marshall Islands will be flooded.

PRG: Do you think this is true?

X: [Pauses] Well, I look at it from the perspective of science, but I also look at it from the perspective of Christianity. God made many kinds of physical features in the world, such as low-lying coral atolls. All of these have their advantages and disadvantages. For instance, you have earthquakes in California, and tsunamis in Japan. But you don't get those here.

I'm not sure about scientists. For instance, sometimes they say, at the weather station, there's going to be a drought, but then it doesn't happen. . . . They say there's a typhoon and then there's not. . . . [But] on Jaluit, where I spent six years, I see coral bleaching. And in Majuro, I also see it. The coral is the color of that wall [he gestures to a dull grey wall]—dead. And in Jaluit, there's no Styrofoam or trash in the ocean, but the bleaching happens anyway. So I think it might be because of global warming. And there's erosion on Jaluit. At high tide at one house, the water gets so close to the house that you could put a chair next to the window, sit on it, open the window, throw a line out, and go fishing from your house! This has all happened since ten or fifteen years go. The faculty housing is right next to the sea. Whose idea was it to build it so close?

PRG: Why is this erosion happening?

X: It could be sea level rise. The North and South Pole—a photo shows how it's getting hotter. There's sea level rise and global warming there. In the past, it wasn't so hot here in the Marshall Islands. Now it's hotter. It used to be just warm. There used to be no droughts. There used to be arrowroot, but now it's gone.

PRG: Why is arrowroot gone?

X: I think maybe because it can't survive above a certain temperature. The sun heats up the dirt. Even at night, the soil is still hot. Look at those coconut trees right there. You see, the fronds aren't pure green—they are yellowish and brownish. The coconuts there are small.

PRG: Because of the sun?

X: Yes.

This man harbors some skepticism of climate change based on reception and exegesis: the image of *scientists as overconfident braggarts* becomes interwoven with the theological notion that God intends for the islands to be habitable. But he nonetheless evinces belief in the threat based on local observation.

A THEOLOGICAL DEBATE

Three Marshallese Internet users, on the Marshall Islands-devoted website Rimajol.com, 2007:

A: Guys, what do you think about the global warming?

B: In front of God, intelligence is stupidity. Don't forget these signs, the end of the world is near. If you read the Bible then you will understand why it is this way. The scientists will offer much of their knowledge but not 100% of what they say comes true [*kūrm̧ool*]. The kingdom of heaven is near.

C: He also gave intelligence so that people could use it to help each other. . . . Remember that he gave you intelligence so you can know what is good and what is bad. There are many among the scientists that also know about the words of the Bible and really believe it. . . . And you Mr. [A], what do you think about global warming and the wisdom of scientists?

A: My understanding is that scientists—not all of them, but those who believe that we came from monkeys—confuse the thought

> of everyone in the world and now they are also trying to find a way to confuse people about global warming, and lots of people will turn away from the wisdom in the Bible.

Here the argument is entirely on theological grounds (exegesis), but a number of views of scientists (reception) are at stake: *scientists as overconfident braggarts*, *scientists as tricksters, scientists as God's rivals*, and *scientists as God's chosen ones.*

The Torn Blanket

Young man, Mejit Island, 2009; a team of American chemical oceanographers from the University of Washington were currently on the island to collect sediment samples to measure past rainfall:

PRG: The visiting scientists are learning about the *mejatoto* [climate, air, atmosphere, cosmos]. What do you think? Has the *mejatoto* changed since the past, or has it stayed the same?

X: There are large changes. Some of the coral has died on the reef. The sun is stronger than before. And you see that breadfruit tree there? It has died from the sun. The coconut trees, too, are brown, because there is little rainfall.

PRG: Why are these changes happening?

X: *You* know why, don't you?

PRG: Scientists say that the world is getting warmer. They call it climate change.

X: Mmmm [a sign of assent]. Yes, I've heard that, on the radio—on the BBC world news. They say there's a layer up high, a blanket [*kọọj*]. It's torn now. So the ice at the North Pole is melting, making the sea rise.

PRG: What will happen?

X: It will cover the islands.

PRG: What will Marshallese people do when that happens?

X: Float! [Laughs heartily]

This man's reasoning combines reception and observation. Like many Marshallese, he used observation to describe the impacts and reception to pinpoint the causes, and uses combination of both to infer the future consequences.

Impacts Everywhere

Middle-aged, well-educated woman, Majuro, 2009. She had attended the WUTMI climate change workshop in April 2009 and had also heard about global warming through her government job.

X: There is a concern. My husband [and I], we were going to build our son's house on a piece of land that we [rented] from some landowners here, and two years ago the shoreline was way down there. Nowadays there is this big tree, it has fallen down. [This is] in Ajeltake [a village in Majuro Atoll].

PRG: When did that start to change?

X: We acquired the land two years ago. In two years, that's really fast. . . . The seasons are changing. The harvests are changing. We used to have *bōb* [pandanus fruit] during December. Now we're having *bōb* season right now [June], which is different from before. So, you know, what is causing all these changes? Is it because the climate is itself changing, it's no longer cool these months, so the breadfruit is adapting to that climate change, it's not bearing fruit where we used to know it's time for breadfruit.

PRG: People used to know when that would happen.

X: Yes, that's something that's very noticeable. We used to know when it's the breadfruit season, when it's the pandanus season. But no longer. Breadfruit season is delayed. Sometimes like where you were in the west [the atolls of Kapin Meto: Ujae, Lae, and Wotho], they hardly have any [breadfruit], because it's so hot the fruit just falls off the trees before they even have a chance to get ripe. And in other places, they almost have them year 'round.

PRG: Which is not the normal way.

X: Yeah. Like on Arno [Atoll], when we were there three months ago, in Longar [Island], that part of Arno, we were eating breadfruit. But when we came to Arno [Island], Arno [Atoll], they said, "Our breadfruits are still not ripe yet." So one side is harvesting, the other side isn't. It didn't use to be like that.

Here observation is primary; but what is being observed is not merely rising temperatures or other identifiable trends but a more general breakdown in predictability—not warming so much as "weirding" (to borrow a phrase from Hunter Lovins).

The Eager Observers

Husband and wife, both middle-aged and well educated, Majuro (normally residing on Ujae and Kwajalein Atolls), 2012:

H: On Ujae, many coconut trees on the shore have fallen. [A neighbor] tried to build a cook house on the ocean side, but it didn't stay. He piled rocks up as a seawall, but it never worked. His cookhouse is gone now—eroded away. The coconut trees that you remember on the ocean side of Arirāeņ [land tract] are gone now too. The same has happened at Lọto [land tract].

W: So this is proof that "climate change" is a fact. . . . Last Christmas the weather was really changing. During the tsunami that hit Japan, the water here went way down. It was just dry land, exposed reef. Then it went way up. This happened in a just a few minutes. It was really windy and rainy that night. So that is really climate change, right?

PRG: This happened because of climate change?

W: Yes. It was my first time to see that—the wind and rain were so strong that it was like a hurricane. Not many saw it because they were sleeping but some saw it. When the tide came back up, it was higher than normal. People were scared and surprised. They thought it was the end of the world [laughs]. The wind destroyed some trees, but just in some spots, not everywhere like in a typhoon—it was strange.

. . .

H: Westerners were the ones who brought the things that create greenhouse gases. In our culture we didn't use fossil fuels.

W: The government isn't really thinking about climate change. You can tell because they are still bringing in fossil fuels.

H: We're talking about fossil fuel, but what about the lasers that they use at Kwajalein [the US military base], and cell phones, and things like that? Cars, trucks, fumes. The government should reduce imports of these things.

PRG: What did you think of the recent high tides?

H: This happens every February. We call it the *iaļap* [spring tide]. In Rita [a neighborhood of Majuro], houses on the ocean side are gone from the *iaļap* that happened in February this year. And big waves washed up over the airport.

W: This happens most Februaries. Old people call it the month of *iaļap*. In February this year the causeway near Ebeye was washed over. Cars couldn't go to Gugeegue.

H: The causeway was washed out so the bus was stuck in the middle of the road. The school shut down for two weeks I think. Those tides were normal, though.

W: But it was more than usual—it washed over the causeway. All the rocks they piled up as a fence are scattered now because of the waves. People said it was because the ocean should be there—there should have been a bridge there, not a causeway, because God made that place to let water flow through. That's how people explained it. They said they should have made a bridge instead of a causeway. That might be part of climate change, right? That people build a landfill in places where the current wants to go. Waves can't go through there so they go somewhere else and erode other parts of the shore.

This interview shows the way in which the scientific concept of global climate change (reception) is becoming an explanatory account for local changes in the Marshall Islands (observation). If anything, the interviewees were *over*attributing local changes to global climate change. At the same time, the interview implies that not every event lends itself to being locally understood as an impact of climate change. Floods are often paraded by foreign concern entrepreneurs as constituting dramatic proof of sea level rise in low-lying countries (Farbotko, 2010). However convincing that kind of evidence may be to western audiences, it is not to Marshall Islanders. I have elsewhere shown that the December 2008 flood in Majuro did nothing to increase local concern about sea level rise or climate change, because flooding events, which have occurred for centuries in this archipelago, are normalized as part of natural seasonal patterns (Rudiak-Gould, in press). A good candidate for being considered a local impact of climate change need not actually be causally related to anthropogenic global warming; it need only seem odd and unprecedented, a perturbation in the cosmos. The eclipse of the sun, the erosion of ancient graveyards, the steady rise of the sea over decades—all of these fit that description, but flooding does not. I will return to these points in Chapter 3.

Mountainless and Luckless

Middle-aged, well-educated man, employee of the government's Marshall Islands Marine Resources Authority (MIMRA), Majuro, 2009:

X: My wife has land in Laura—at the end of the island. We built a shower and little house there, so people could take a shower after they went for a swim. The shower's still okay, but the waves have destroyed so much—they destroyed the little house. It's gone from the erosion.

PRG: Did there used to be erosion in the past?

X: No. It used to be very good. Now it's very bad.

PRG: Why?

X: High tide. Climate change.

PRG: I saw the eroded cemetery in Jenrōk.

X: Yes, it's gone. There used to be many graves there. Then it was severely damaged. The land used to go much further out. That's not the only cemetery that this has happened to. There's one in Uliga. There used to be lots of graves there too. Now it's all gone.

PRG: Why did that happen?

X: The ocean is higher. It wasn't that way before—it was very good.

PRG: Why is climate change happening?

X: The ozone is broken, so it is sunnier now. . . . They emit things into the atmosphere. I don't understand it too well. It's not just

> America but all countries . . . damaging the *mejatoto*, making the sun stronger, so the ice is melting. The Marshall Islands will disappear. Lucky for the Federated States of Micronesia that they have big mountains—there are none here in the Marshall Islands.

He speaks of observation, specifically the erosion of distinctive and culturally valuable sites, while his description of the causes is guided by reception—his exposure to scientific narratives of global climate change.

Unaware and Unconcerned

I administered my survey in 2009 to an eighty-three-year-old woman living in Delap, Majuro. She had spent all but the last few years of her life in the outer islands. Although she had completed some high school, she had never heard of the scientific idea of climate change. She reported no changes at all in the weather, climate, or ocean and said words to the effect that the Majuro flood in December 2008 and the erosion of the graveyard in Jenrōk were equally common in the past.

When I was conducting fieldwork on Ujae Island in 2007, there were a number of individuals who had nothing to say either about the scientific idea of climate change or about local impacts, even though I knew them well and they were quite happy to speak to me about many other topics, including unpleasant ones. On Ujae, I noticed that a number of coastal trees had fallen since I had last lived on the island in 2004. I asked locals about this, wondering if it would elicit comments on local environmental change or on the scientific idea of global warming. Very often it did not. Sometimes the individual would imply that nothing was amiss—that only the one tree had fallen, that the tree was simply old, or that it had been knocked down by the wind—and then have nothing else to say on the matter. One time I sat with a local schoolteacher on the very roots of an eroding tree—a large *lukweej* (*Calophyllum inophyllum*) tree, certainly quite old, whose roots were half exposed from erosion and which seemed destined to fall in a few years—in a spot surrounded by already fallen pandanus trees, and we talked about the shipwrecked fishing boat that had sat on the reef for many years and had recently begun to decay rapidly from wave action. Despite all of these reminders of erosion and sea level rise, when I asked the man why the grounded boat was eroding, he simply said, "I don't know," looked momentarily uncomfortable, and then regained his poise and said nothing more on the topic. At the time, and when I wrote about this previously (Rudiak-Gould, 2009a), I interpreted his response as the deliberate avoidance of an uncomfortable issue or even an implicit denial of its reality. Perhaps so, but it may also be that these individuals were simply unaware of climate change by any of the channels of information I have discussed. Naturally this book focuses on the majority of Marshall Islanders who *are* aware of climate change in various ways, but it is important to note that people do not always know or care about what we call climate change, an important fact sometimes unmentioned in commentary on indigenous attitudes toward global warming (see for example Johnson & Levin, 2009: 1601; Nilsson, 2008; Parker et al., 2006).

The Threatened Church

Middle-aged, well-educated woman, involved in education and a women's organization, Majuro, 2009:

X: You see signs of [climate change]. [But] in terms of sea level rising and covering all these islands, I don't know. I believe that it will happen in the future. But I don't know. . . . Some people believe that it won't happen.

[On] this one island where I grew up [Imrooj Island, Jaluit Atoll], we have a church. Last time I went there, the shoreline came all the way close to the church, to about maybe 15 feet from the church. Whereas when I was growing up, it was way out there. But now they're saying that the shoreline is growing again, coming back. . . . What does this tell us in terms of climate change and erosion? . . . What does it mean? And of course you want to believe that maybe it's not happening. And then you see that kind of sign and you [think], "Oh, what does it mean?" . . . I'm sure there is some scientific explanation. But I'm not a scientist so I don't know why that happens. But I like to think that it's not happening. But of course I read all of these other reports and explanations, so I know that it's also happening. So I guess you can never be 100% sure, which is why you still need to prepare for it.

This thing on Imrooj was really fascinating for me because I lived there till I was about twelve years old. So my formative years were there. My father was a minister in that church, and we lived in that place. And so when I was in eighth grade we moved back here. And I went to Rita [a neighborhood of Majuro] and then went on to high school. . . . After I came back from college, and I was working with the Ministry of Education, I went back there [to Imrooj], and I was like, "What happened? Did they move the church to the shoreline?" And they said, "No, the shoreline came to the church!" [Laughs] Really, I was saddened by that. Because I was thinking eventually it's going to come all the way and the church is not going to be there. Then, you know, this thing happened and I'm like, "I need to see for myself.". . . These people were like, "Oh! You wouldn't believe it but the shoreline is being built up again and it's extended out there and the threat to the church is not there anymore."

PRG: When I ask people about the [December 2008] flood, some people say, "It was worse than the ones we had before, so it must be climate change." But other people say, "Oh, it's just *iaḷap* [spring tide]. It's that time of year. It has happened before, and other ones in the past were worse."

X: Yeah, yeah. You see different responses. Sometimes I wonder if we were not exposed to the global discussions about climate change, would we have. . . . Because certainly we're influenced by that. We read all these things and then you see a little thing and you say, "Oh! It might be [climate change]." So I don't know, to tell you the truth, I don't know, but I like to think that the scientists know what they're talking about.

But you know there was another island that belonged to my family on the southern part of Jaluit. It's called Piñḷap. . . . It's a pretty big island. And lush, lots of food, different kinds of things grow really well. But on the western side of the island, there's a lot of erosion. Erosion has really taken its toll. There were graves on that part. They're all falling to the ocean side. . . . Their families belonged to my families a long time ago. . . . So I'm like, "Wow. This is really . . . this whole thing on climate change might be really [true]." And I'd like to see—will that continue to happen? Or whether at a future date something's going to change, the current will change and bring back the sand to that side of the island. . . . So there is this side of me that wants to believe that climate change is not going to happen, because of course I want to know that when I die that the Marshall Islands will still be there. [Laughs heartily] That's what I'd like to die knowing—that they're safe and they'll be there, but who knows. So there is that side of you that wants to believe that this climate change is just something that's temporary.

While tending toward trust of scientists and their climatic prophecy, this woman feels that observation leaves some room for doubt. She also frankly admits that observation is influenced by reception of scientific discourses: "Sometimes I wonder if we were not exposed to the global discussions about climate change, would we have [noticed local impacts]. . . . Because certainly we're influenced by that. We read all these things and then you see a little thing and you say, "Oh! It might be [climate change]."

Visions of Destruction

Elderly woman, Majuro, 2009, after she attended the WUTMI Executive Board Meeting:

X: Climate change. Global warming. Warming. I had a vision yesterday, in my mind, when they were talking about climate change at the conference. This is the fourth time I've had this vision. It is a vision of global warming. I've had other visions too. I saw the Kwajalein strike of 1982 in my mind before it happened. And the evacuation [of the Rongelapese] to Mejatto—I had a vision of that before it happened. Now I see global warming in my head. The vision is like this: There's an island, and the ocean is eroding it more and more. The island is getting smaller and smaller. There's a tree on the island. It's dead with no leaves—dead from the hot sun, and from the warm wind. Then the wind comes, and the tree is tilting over and almost falling. The wind is global warming—global warming is like a strong wind. The climate change will erode the islands, and the global warming will kill the plants with warm winds. It's not a cool wind like the kind we have these days. We'll be dead before climate change happens. *You* will live to see the damages, but we'll be dead.

PRG: When was the first time you saw this vision of climate change?

X: In 2002. The first vision was of an island with many trees, and the sea starting to erode the land. Then I had the vision of the tree and the wind, the one I just described to you. . . . They say it's time for the world to end, in the Bible. . . . In the time of Noah, the waves rose above the mountains. God said, "I will not destroy the world again with a flood. I will destroy the world with brimstone and fire from heaven." Brimstone is like the warming [from climate change]. People have damaged the world, but God loves us still.

PRG: When will the world end?

X: I don't know. But it looks like my vision of climate change is part of it. There will be famine. *You* will see it, but I'll be dead by then.

PRG: So you think climate change is part of the end of the world?

X: Yes, I think so. I'm not like those who didn't believe Noah.

PRG: Do you believe what scientists say about climate change?

X: Yes, I must trust them. They didn't use to see changes, but now scientists do. The scientists see the changes. But, scientists shouldn't do things like fix or transplant people's hearts and eyes—that's *God's* work!

PRG: Do you believe in climate change because you trust scientists?

X: No, I believe because I trust the Bible and God.

She relies on both reception and exegesis. In addition, she uses her own idiosyncratic source of information: supernatural premonitions.

Human-Caused Harms

Two women, one young [Y], one middle-aged [X], Majuro, 2009. Both work at Assumption School; one used to teach Bible studies at the school. Both had attended the WUTMI climate change forum in April 2009.

PRG: What did they talk about at the WUTMI forum?

X: They talked about how the *mejatoto* [climate, air, atmosphere, cosmos] is changing—how people are causing this. . . . We learned that Styrofoam cups are bad. They were brought here from outside the country—they didn't come from here. People discard them and this is bad. So I decided to do my part. I noticed that at the forum, while they were talking about how Styrofoam was bad, one of the women was drinking from a Styrofoam cup! So I raised my hand and said, "You shouldn't use that"—so they stopped. I was very direct [*kajju*] about it. But there was another woman who kept drinking from a Styrofoam cup. [Laughs.] Styrofoam ends up in the ocean and damages coral. In the past there was nothing like Styrofoam in the Marshall Islands—it came from *other* countries. . . . There was also a man from the College of the Marshall Islands who told us about how stuff goes up into the sky and heats up the world. So we should really conserve gas and use cars less—really conserve gas. The climate changes, the ice melts, it gets hotter, and low islands are flooded. . . . A big problem right now is that the country is polluted. We need to stop this.

PRG: What will happen in the future?

X: The *mejatoto* will change, these islands will be covered and then there will be nothing. Illness will come. We have too many things from outsiders—like money. We don't rely on our own things anymore. We depend on outside things. We don't grow our own food anymore.

PRG: Do you see changes from climate change already, now?

X: That is funny, because just yesterday, [Y] and I were sitting just here at Assumption, and looking at the ocean side, and we saw that it was extremely low tide. There was no water at all, just exposed reef. That's not like the tides in the past. In the old days, people would say, *Añōneañ*, *rak*, but not now. They could expect that December and January would be the times of big waves, but not now. This change is due to climate change. You get very low tides.

Y: It used to be that the summer was hot, with hardly any wind, and then in Christmas and January or February, it was called *añoneañ* and we expected storms during this time. But now it has changed. And it's getting really hot now.

X: And you see erosion nowadays too, on the small islands . . . Bikirin and Enemanit [islets in Majuro Atoll], and the end of Laura. The end of Laura has disappeared. . . . God gave us these small islands. They were a gift from God. We had no money but we were blessed with coconuts, breadfruit, and pandanus. . . . In the old days, on Jaluit, when you made food you would bring a serving of it to each house on the island. Now, when I went back there, they don't do it very much—soon it will be gone. It's weakened, because of western influence. In America, when a child reaches eighteen years old, they leave the house. That's not how you do it in the Marshall Islands. My son lives in Missouri, with an American wife. He says that when his children are eighteen, they will leave. That's how American culture is. You only take care of yourself. When a problem comes, you can only rely on yourself to help you.

PRG: Who is saying that climate change will happen? It's scientists, right?

X: Yes, the clever people [*rimālōtlōt*].

PRG: So you trust scientists?

X: Yes. Just a few days ago, on CNN they were saying that there would be famine in the world in the years to come. I mentioned this to another woman, and she got really angry. She said, "Don't listen to what CNN says. God will not destroy the world with all kinds of things." [Laughs heartily] But she was wrong, because it's *people* who are the ones harming the world.

PRG: I've met people who say, "It won't happen because God promised he wouldn't destroy the world again."

Y: It's true what God said, but it's people who are doing it. We're using our intelligence to destroy the world. We're the ones not keeping our side of the promise.

X: Yes, that's true.

Using reception, observation (including an atypical report of lower tides), and exegesis, both find they are entirely convinced that climate change is real.

Global Warming versus God

Middle-aged man, Majuro, 2009. I talked to him at low tide on the ocean side of Delap, Majuro, while he was fishing at the *bōran baal*, the outer edge of the reef. Many seawalls along the shore were visible from where we were standing, and I asked him about them.

PRG: Why did people build these seawalls?

X: Because the waves rise up and cause damage. You see that damaged house there? Waves did that.

PRG: Did this happen in the past?

X: No, only now. I left Majuro in 1988 and lived in Oregon for sixteen years. When I came back, things had really changed. Now the ocean is high. Each year I see more changes happening.

PRG: Why is the ocean higher nowadays?

X: Well, according to the clever people [*rimālōtlōt*], the ice in the cold parts of the world, like Alaska, is melting and creating water.

PRG: Do you think that's true?

X: [Hesitates] A bit, because I see that the tides are higher nowadays. But it's obvious that God protects us here because there are typhoons that really cause a lot of damage to Guam, but they don't cause as much damage here in the Marshall Islands.

PRG: Do you think the *mejatoto* [climate, air, atmosphere, cosmos] has changed since the past, or has it stayed the same?

X: It's hotter now. There's less rain. There are many changes. Some corals are dead.

PRG: Have the waves changed, or have they stayed the same?

X: It's not like before. Now the waves reach to the height of the land.

He wields observation, reception, and exegesis; he trusts scientists (calling them "clever people") and sees local environmental change, but exegesis gives him room for doubt.

Only Anecdotal Evidence

Young, well-educated man, Majuro, 2009. Extensively educated abroad, he certainly believed wholeheartedly in climate change, having written a letter a few years previous to then-president Kessai Note, advocating a more proactive stance on the issue. He was also deeply convinced that the country was deteriorating more generally, as he had argued in a commissioned report to the government. Nonetheless, he expressed skepticism about whether local environmental change was truly occurring and about whether purported changes could be attributed to global warming.

> In Tuvalu and the Solomon Islands, there have been very dramatic impacts of climate change, so it grabbed their attention. But not here in the Marshall Islands. We have some small anecdotal stuff, like that Mile 17 [a well-known beach in Majuro Atoll] has been eroded, but not enough for people to care. And here in Majuro, it's hard to tell between what has been caused by dredging versus what has been caused by natural processes. And within the category of natural-caused effects, is it climate change or some natural cyclical thing? I haven't made up my mind about it. Like the big erosion at Laura beach—is that caused by dredging, or a seawall, or climate change, or a cyclical process? And erosion on the outer islands, is it climate change or not? We need long-term data to know, but we only have anecdotal evidence. Ailuk [Atoll] people say there's lots of erosion there. They had to get grant assistance to do a revetment there, because the erosion was affecting the runway. But I would be cautious to say it's because of climate change.

For this man, reception was clearly primary and conclusive and observation secondary and inconclusive. Quite explicitly, he had heard of climate change first, come to accept its reality based on the science, and only then looked for signs of it in the local environment.

The Trillion-Dollar Problem

Elderly woman, Majuro, 2009. She had attended the WUTMI forum on climate change in April.

> I live in Laura, and a lot of it has disappeared from the ocean. The pandanus trees are gone. There used to be coconut trees, but they've fallen from erosion. The waves are eroding the whole shoreline. . . . I can confirm it. The wind is strong. Not like in the past. It has started to rain less. They say there's an island on Utrik Atoll where it hasn't rained in two years. They had to bring in a machine to make fresh water. It didn't used to be this way. Now when I go to Laura and look at the water, I can see that the water is rising. It's very clear that the water has risen. Perhaps soon it will reach the level of the road. I see it and I say, "It is very clear." . . . The islands are thinner now. I don't know what the government is going to do. The RMI has no money. It might take a trillion dollars. Not just in Majuro but in outer islands too there are problems. It's happening very much, from what they call *oktak in mejatoto* ("climate change"). . . . On Utrik Atoll, I've heard that there are few breadfruit now. There used to be lots. And all the coconut trees have died. We see what the scientists are saying. Now in Laura, there are wells that used to have lots of water. It would never run out. But now it does. And I think this is part of climate change too. . . . Sometimes coconut trees and banana trees are damaged by the wind, even though there's no typhoon. Sometimes the sound of the wind is really loud. Sometimes after just four hours of wind, a tree falls down. And I think that's probably also part of climate change because it comes from the *mejatoto*.

Here both reception and observation come into play, with observation serving as confirmation of reception.

The Hyperbeliever

Elderly, well-educated man, Majuro, 2009. He had been involved in the government for decades.

> Just yesterday, NOAA gave a warning—they said not to expect any rain till July in the North Pacific. But today [June 21], it's really rainy!

> So even *they* don't know what's going to happen—that's how screwed up the seasons are now.
>
> See that island there? [Points to a small island.] Those trees there have been eaten away and are falling down. . . . It's June now, and it's dry. In my sixty-five years in the Marshall Islands, I've never seen a dry summer. And I've seen an island disappear in Majuro lagoon. It's the one just south of Kalalin. That island is gone like you dropped an A-bomb on it. Water now comes to where my mother used to grow flowers. On the small islands of Majuro, you'll see what I consider to be an extraordinary algal bloom—there's tons of seaweed growing. These blooms used to just come and go in a matter of days to weeks, but now they stay for twelve to eighteen months. It's spooky. [He shows me the seaweed growing in the very shallow water in the lagoon next to his house.] It's not too strange if it's here in D-U-D, because we know how environmentally damaged this area is. But if it's out on the small islands of Majuro, which should be more pristine, that is scary.
>
> Mile 17 [a well-known beach] has lost all of its beach sand. Tides are higher now. Fish species are changing too. Now we have saltwater catfish. We've never seen them before. They have no name in Marshallese. And more whales are being seen, off the ocean side—and one time recently a whale even came into the lagoon. And there are new bird species, because of climate change. Everything is changing. This is supposed to be the rainy season right now, but it's dry!
>
> Jacques Cousteau gave a warning about all of this at the 1991 Rio meeting. What he said then is happening now, exactly as he predicted. At the time, he was looked on as a naysayer, someone who was exaggerating. But he was right.

This man is so confident that climate change is real that even an example of scientific prediction gone awry is taken as yet more confirmation that the world has changed beyond recognition and that climate change is therefore real. Strongly convinced by reception, his observation becomes equally confident, to such an extent that he sees signs of climate change in nearly everything. Such observations then strengthen his confidence in the scientific theory, which in turn encourages him to see more signs of it, and so forth, so that reception and observation form a feedback loop and the man is ever more convinced of both.

The Occasional Skeptic

I talked many times with an elderly man whom I knew from Ujae in 2003–2004 and who was living in Majuro during my 2007 and 2009 fieldwork. When I asked him about climate change in August 2007, his triangulation left him unsure. He saw environmental change in the form of waves exposing ancient graves on Majuro and Ujae, as well as the erosion of beaches down to the reef platform on several uninhabited islets of Ujae Atoll beginning in his boyhood. He had also heard the scientific prediction of devastating sea level rise on the radio when the Nitijela discussed it. But he felt that God's promise might outweigh both his own observations and the proclamations of scientists. When I spent time with him again in May 2009, this uncertainty had manifested itself as either a confident dismissal of climate change or a confident affirmation of it, depending on the day and his mood. On May 14 he rejected the idea of climate change:

PRG: Did you experience the flood last December?

X: Yes. It was here around my house.

PRG: Did it cause any damage?

X: No. It didn't enter the house except for a little bit near the door. On Ujae, there was flooding, but not much. There was no damage. . . . Each year you see the *kapiḷak* high tide. The flood last December was the *kapiḷak*.

PRG: So it was the same as other years?

X: Yes . . . God makes it so. It won't change. It will always be the same.

PRG: Some say that the sea is rising.

X: God made a promise that the sea won't rise again. The rainbow is a sign of that promise.

PRG: So scientists are wrong?

X: Scientists become arrogant. They want to have Godlike knowledge. They helped people to get to the moon. But God didn't put people on the moon. He put them on the earth. They aren't supposed to be on the moon. They want to know more than God. But they don't know when they're going to die—only God knows that. They're so smart that their thinking changes. They say that there is no God, that people came from animals.

PRG: Some people say that sea level rise is a punishment for Marshallese people's sins.

X: You mean like the Flood in the Bible? No. Because God promised he wouldn't flood the earth again. Only once. [But it's true that] the last day will come. Sons will fight fathers. They will be hunger and thirst. Everyone will see God—he will be sitting there in the sky, wearing a crown. People in every country will see him, and people who don't believe will go to Hell for thousands of years.

PRG: When will this happen?

X: No one knows when. [Smiles] Scientists don't know either!

Here the extent of local environmental change is minimized, the reliability of scientific knowledge is impugned (using the *scientists as God's rivals* idea), and God's supreme knowledge is affirmed. But just nine days later, he offered a different view:

PRG: Why did the waves erode the land at Mijijak land tract?

X: They say that the ice is melting, in Greenland. It's hotter nowadays in the world. Americans are causing this. They fly up into the sky, breaking it. That's why it's easier now for the waves to erode the island. The land slides down at Bateṇ land tract. You used to not be able to see from Ariräeṇ land tract to Lọākā land tract [on the other end of the lagoon-side bay], because there was land blocking the view. But the waves took it away. There was probably fifty feet more land there—fifty feet farther out into the lagoon than now. Now it's gone—at Ṃōnluni land tract. Coconut trees are falling down there, and pandanus trees. Americans flew up to the moon. This changed things.

On Ebeju Island [in Ujae Atoll], there are very old graves—people from the distant past. . . . Maybe American scientists can come and get information about these graves—figure out if the bodies are men or women and how many years ago they were buried.

The *bar* [rocky shoreline areas] protect the island, making it hard for waves to erode the island. But now people take sand and rock and break it, and they dig on the lagoon shore in order to build seawalls, so now it's easy for waves to erode the island. This is happening in Majuro. They didn't do this in the past. Soon the erosion on Ujae will reach all the way to the road. Scientists say the ocean is rising.

Tradition is fading. Marshallese food is gone. Arrowroot doesn't grow anymore, maybe because of the bombs. . . . They should grow by themselves, without being tended. . . . But nowadays they get about a foot or two high and then die.

PRG: Do you think what scientists say about sea level rise is true?

X: The ice melts, making the ocean higher. It's probably true.

The man's ambivalent opinion of climate change seems to stem from the ambiguity of both observation and scientific expertise. In the first conversation he downplays local environmental change, declaring the December 2008 flood to be minor and natural, whereas in the second conversation he points to large changes in both the city of Majuro and the rural island of Ujae. In both conversations the idea that scientists/Americans flew to the moon comes into play, but with opposite entailments: first it is an example of *scientists as God's rivals*, proving scientists' untrustworthiness and therefore the falsity of climate change; then it is claimed to be a *cause* of climate change, proving its truth, and scientific expertise is flattered in the man's suggestion that scientists could uncover reliable information about

Ujae's ancient graves. It is unlikely that in those nine intervening days the man had changed his mind about so momentous an issue as climate change. More likely he harbored both views, and claimed one depending upon the specifics of his triangulation at the moment. His mood may have made a difference: in the second conversation we had been talking about the changes he had experienced since his youth, the loss of *m̧antin m̧ajeļ*, predisposing him, perhaps, to pay attention to negative environmental change.

In 2012, this man passed away after a long fight with diabetes, and was buried on his ancestral island, Ujae. That grave may one day meet the same fate as those he had watched disappear around him. This book is dedicated to him, in the hopes that somewhere in the uncertainties of the changing climate—uncertainties that he captured so well—might lie the possibility of a long and undisturbed rest.

3 Pervasive Decline and the Eminent Believability of Climate Change

Many activist and scholarly accounts present indigenous communities as alarmed witnesses to change, united in their conviction that dangerous flux is afoot (Parker et al., 2006), while westerners are portrayed as unconcerned or skeptical of climate change (Norgaard, 2006; Stoll-Kleemann, O'Riordan, & Jaeger, 2001). The assertion is not false: indigenous peoples everywhere are reporting strange climatic shifts and many communities are taking action (see, for instance, Crate & Nuttall, 2009b), while skepticism persists in the West, and northern governments and citizens have done far less to address the crisis than they might have. Still, this narrative is limiting for at least two reasons: it assumes that only *disbelief* and *lack of concern* about climate change need be explained—as if the objective truth of climate change were enough of an explanation for why people accept its reality—and it assumes that frontliners believe in climate change solely because they observe local changes in weather patterns and "environmental" conditions.

I hope that I have already refuted, through my discussion of triangulation, the assumption that local observation is all that matters for belief. But the scope of evidence that can count for or against the reality of climate change is actually even wider than the previous chapter indicates. Belief can hinge on observation of phenomena which, from a western or scientific perspective, have nothing at all to do with climate change.

On July 22, 2009, four men were waiting on the leeward shore of Leb Island for the coming of *marok eo*, the darkening of the sun predicted by scientists on the radio. As daylight turned to twilight and the sun turned black, the men were amazed. Neither they nor anyone else on the island had witnessed a total solar eclipse before. Some locals had gone to church; the church bell had been rung to announce the eclipse. Other people stayed indoors, frightened by the strangeness of the event, and prayed to God. Some saw the event as a possible sign of the coming biblical end-time. A young man told me, "In the Bible it says that when the last day comes, God will perform miracles [*men in kabbwilōñlōñ*]. Well, this darkness was a miracle."

Life quickly returned to normal afterwards, but the eclipse had left an impression on the people of Leb Island and many other Marshall Islanders. Indeed, it had altered people's attitudes toward the threat of climate change. It was not simply that the event had influenced islanders' trust of scientific predictions, as discussed in the previous chapter. It was also that the event itself had been considered linked to climate change. After the eclipse, when I asked people on Leb Island, Mejit Island, and Majuro if the climate (*mejatoto*) had changed or stayed the same since the past, many answered, "It has changed—the sun went dark." The eclipse was thus being cited as evidence for, and in fact an example of, climate change. The eclipse had perhaps done as much as all of the year's outreach efforts to convince people that scientific prophecies of climate change were to be trusted. For locals, climate change was not a new and singular phenomenon but part and parcel of a general transformation already under way.

CLIMATE CHANGE BELIEF AND MODERNITY THE TRICKSTER

While the previous chapter presented some Marshallese skeptics on climate change and their reasons for skepticism, by and large Marshall Islanders believe[1] in climate change. Of the eighty-nine Majuro survey respondents who had some familiarity with the scientific concept of climate change, 67% said it was definitely true, 17% said it was probably true, 11% said it may or may not be true, and only 4% said it was false; thus 84% were more or less convinced that it was true while only 4% were convinced that it was not. In the Marshall Islands there is no organized or public opposition to climate change. A variety of initiatives—including awareness raising (Chapter 2 and later in this chapter), mitigation (Chapter 4), and adaptation (Chapter 5)—are all pursued on the premise that climate change is real and human-made. Acceptance is the safe, uncontroversial position.

Concern, though not panic, is also widespread. Concern, in the sense of preoccupation, can be fruitfully measured by counting how often people mention a particular threat when asked very general questions about what problems exist in their society. Research shows that people have a "finite pool of worry" (Slovic et al., 1987: 26; Weber, 2006: 114–115): with limited cognitive and affective resources for being concerned, preoccupation with one issue must come at the cost of preoccupation with another. I therefore calculate the concern of an individual by dividing the number of mentions of climate change by the total number of mentions of all issues by that individual. (For instance, if a respondent mentioned the issue of climate change once and the issue of financial hardship twice, his or her climate change concern would be calculated as 0.333.) I then calculate the mean of this measure for all respondents. Here I use only the first batch of the 2009 Majuro survey (with 100 respondents), in which I asked such questions *first*, before asking about the scientific concept of climate change or about the respondent's observations of environmental change—so the respondents were not primed by a previous mention of climate change. The results of this calculation are presented in Table 3.1.

Table 3.1 Ranked Concerns among Majuro Survey Respondents

Rank	Issue	Concern Measure (rounded)
1	Economic hardship and basic needs	0.25
2	Changing lifestyles and mores	0.21
3	Population growth and overcrowding	0.08
4	Diabetes and other health problems	0.07
5	Changes in the climate, changes in the ocean, "climate change," "global warming"	0.06
6	Alcoholism and other substance abuse issues	0.06
7	Immigration	0.06
8	Crime	0.03
9	Education	0.03
10	Suicide	0.02
11	Youth problems in general	0.02
12	Teenage pregnancy	0.01
13	Outmigration	0.01

In this rough estimate, climate change emerges as the number-five concern, edged out by economic, cultural, demographic, and medical worries but trumping such issues as substance abuse, immigration, suicide, and youth problems. I consider this an impressively high level of concern given the following facts. Climate change is a notoriously difficult hazard to muster concern about. There are a host of psychological barriers that prevent people from protecting themselves from hazards, such as the possibility of denial (see, for instance, Edelstein et al., 1989; Lehman & Taylor, 1987) and the related optimistic bias (Weinstein, 1989: 44) and positive illusions (Johnson & Levin, 2009: 1596) by which people tend to believe themselves to be less vulnerable than others around them. In addition to these general barriers, climate change presents special liabilities (Ungar, 2007: 85). Its impacts are uncertain and lie mainly in the future, falling prey to what psychologists and behavioral economists call future discounting (Frederick et al., 2002), or the tendency to care less about future harms than current ones. According to the "Giddens paradox," the impacts of climate change become clear only once it is too late to prevent them (Giddens, 2009). Present effects of climate change are not always dramatic (Sunstein, 2006a; Weber, 2006), and presentations of pallid, statistical scientific information may fail to engage people emotionally (Marx et al., 2007) and meaningfully (Jasanoff, 2010), even in highly vulnerable societies (Patt & Schröter,

2007: 8). There is no "9/11 for climate change" (Sunstein, 2006b: 4), even in a vulnerable nation like the Marshall Islands (Rudiak-Gould, in press). Individual events are extremely difficult to definitively attribute to global climate change, since any environmental disaster has multiple causes and global warming manifests itself in broad, worldwide trends unfolding over decades rather than in discrete one-off events. Relatedly, the causation of the hazard is difficult for those without scientific training to grasp (Sterman, 2011; Ungar, 2007: 83), potentially reducing concern (Bostrom & Lashof, 2007: 38–39). The legal scholar Cass Sunstein has contrasted these troublesome characteristics of global warming with the far more straightforward attributes of terrorism—9/11 happened suddenly rather than gradually and had an easily identifiable cause and perpetrator—to explain why the United States government has been willing to invest vast amounts of resources to prevent future terrorism but not to prevent future climate change (Sunstein, 2006b). The finite pool of worry means that climate change must compete for concern with a host of other dangers, many of them more harmful at present than global warming (Hezel, 2009a). For the impoverished and marginalized, those who are most climate-vulnerable, the pool of worry is already full to overflowing. The people most endangered by climate change may have the least luxury to devote thought to it.

This helps to explain why concern about climate change is low in many societies. In the 2006 Nielsen Global Omnibus survey of climate change perceptions in a variety of large nations, in every surveyed country only a tiny percentage of respondents said that global warming was their first or second concern over the next six months (Boykoff & Roberts, 2007). Even in some communities severely vulnerable to climate change, and where large changes have already been observed, such as in the Arctic (Bravo, 2009: 277; Fox, 2002: 45; Huntington et al., 2005: 62), Tuvalu (Mortreux & Barnett, 2009: 110), the Himalayas (Byg & Salick, 2009: 164), and Arnhem Land, Australia (Petheram et al., 2010), many other issues are discussed far more often than climate change. Climate change "has some real liabilities as a marketable social problem" (Ungar, 2007: 85).

Keeping in mind the finite pool of worry, consider the many problems apart from climate change with which Marshall Islanders must preoccupy themselves. A 2006 government-administered study revealed a host of local worries such as alcohol and drug abuse, crime, unemployment, teenage pregnancy, overpopulation, unreliable electricity, poor transportation, and food shortages (RMI Government, 2006b). In 2006, an estimated 20% of the population lived on less than US$1 per day (Juumemmej, 2006: xiv). Forty percent of schoolchildren are estimated to be malnourished (Asian Development Bank, 1997: 102). Suicide rates are high and increasing, linked to rapid social change (Booth, 1999). Domestic violence (Journal, 2003b) and other forms of violence (Carucci, 1990; Johnson, 2004) are far from uncommon. On Ujae, where I lived for a year, locals had to cope with low cash incomes (reportedly a mere US$204 per year in 2006 [Juumemmej,

2006: 20]), periodic food shortages, the possibility of droughts in the dry season, insufficient sailing canoes, the high cost of transportation to Majuro and Ebeye, the infrequency of the supply ship and unreliability of the airplane, and the unavailability of certain medicines, not to mention more pedestrian concerns about wayward children and feuding neighbors. According to a well-researched socioeconomic report, "All the major indicators suggest that the quality of life for most Marshallese is worsening" (Juumemmej, 2006: 137) for reasons unrelated to climate change.

Given this, it is striking that climate change is nonetheless people's number-five concern, ahead of such immediate and pressing issues as substance abuse (including alcoholism, a major problem in the Marshall Islands), immigration (including the recent influx of Chinese people, which is the source of much resentment and frequent discussion), youth problems, and nuclear testing. A few locals confessed to me disturbing dreams about climate change: seawalls being constructed higher and higher until Majuro was encased in concrete. Some students on Wotje Atoll, having recently heard about climate change for the first time from an American expatriate teacher, complained of sleeplessness as they listened to waves crash on the shore. Although few if any Marshall Islanders are experiencing what has been dubbed "ecoanxiety"—obsessive worry about coming environmental disasters (Nobel, 2007)—and no one is visibly panicked about climate change, concern is impressively high.

It is climate change *skepticism* and *disbelief* that scholars fixate on (see for instance Besnier, 2004; Brugger, 2010; Donner, 2011; Feygina et al., 2010). But Bruno Latour (1993) has warned us against the fallacious assumption that true beliefs are believed because they are true: we must explain people's belief in *true* propositions in the same way that we explain people's belief in false propositions. Why then do Marshall Islanders find climate change plausible? If Marshall Islanders wish to deny the reality of climate change (and it is not hard to see why they would be tempted to), they have plenty of locally credible grounds on which to do so—point to the wily untrustworthiness of scientists or brandish the Genesis argument; the previous chapter showed the many avenues that Marshall Islanders have toward disbelief. One study in the nation of Tuvalu reported that 55% of the informants doubted the scientific prediction of climate change because of God's promise in Genesis (Paton & Fairbairn-Dunlop, 2010: 691). Why is skepticism not similarly rampant in the Marshall Islands?

In this chapter I will argue that a major, if not *the* major, reason that Marshall Islanders by and large accept the truth of this prophecy is that it fits well—uncannily well—with the narrative of modernity the trickster, to which they are already so committed. Traditionalism indeed has much in common with environmentalism. Western environmentalism narrates a fall from a pristine past to a corrupted present and ruinous future (see Easterbrook, 1996; Hulme, 2009: 342–348; Norgaard, 2002). The blame for this decline is understood to be on human beings, who have foolishly embraced new technologies whose seductiveness is matched only by their destructiveness. The former goodness cannot be wholly regained, but people can safeguard what remains by casting aside destructive artifacts and habits. Environmentalism thus evokes feelings of guilt (Swim et al., 2009: 85–86), loss of the past (Hulme, 2009: 342–344), and fear of the future (Douglas & Wildavsky, 1982: 127; Hulme, 2009: 345–348) while also holding out the possibility of redemption. Borrowing religious terminology, Mike Hulme calls the environmentalist narrative of loss "lamenting Eden" and the environmentalist narrative of fear "presaging Apocalypse" (Hulme, 2009: 342–348). Regarding fear and guilt, Mary Douglas and Aaron Wildavsky write that the dominant discourse of "sectarian" societies, in which they include the more radical western environmentalist movements, "expects life in the future to undergo a radical change for the worse. It is not confident that the disaster can be averted. There may be no time left. But it knows how the disaster has been caused: corrupt worldliness" (1982: 127). They continue: "in a secular civilization nature plays the role of grand arbiter of human designs more plausibly than God" (Douglas & Wildavsky, 1982: 127). Anthropogenic climate change perfectly fits this narrative structure: climatic stability, once taken for granted, now slipping away; looming meteorological catastrophe; human (or northern) culpability; the atmosphere and the carbon cycle acting as the "grand arbiter" of human aspirations.

In narrative structure, in emotional resonance, the Marshallese discourse regarding modernity the trickster is identical. The only difference is that, here, the "grand arbiter of human designs" is neither nature (for which Marshall Islanders have no category) nor God (for whom Marshall Islanders have a category but of whom they speak less often than they do of tradition) but *m̧anit*; it is Marshallese custom that acts as the omnibenevolent protector and avenger. The ethnic/spatial (foreign versus local), temporal (past versus present versus future), and moral binaries of traditionalism are all upheld; the idea that a formerly benign, orderly, and correct climate is becoming malignant, chaotic, and incorrect accords perfectly with the same belief regarding culture. Therefore to embrace the scientific discourse of climate change, Marshall Islanders do not need to grasp a wholly new concept but merely to add another binary (good climate:bad climate) to the list. Global anthropogenic warming is a new idea, but for Marshall Islanders its basic structure is so familiar, so taken for granted, that acceptance is an easy position to take.

I documented just one instance in which an informant *explicitly* equated the scientific narrative of climate change with the local narrative of cultural decline and modernity the trickster. The following is from a 2009 interview in a combination of Marshallese and English with a well-educated woman associated with WUTMI:

PRG: Why is climate change happening?

X: Because of all this man-made stuff, like [motor]boats and high buildings . . . I think the climate change is like the custom change. It's the same thing. Because of all these new things.

PRG: Who is making it happen?

X: Us people. Including Marshall Islanders. Because we follow American culture [*ṃantin pālle*]. We make big buildings . . . Before we used to rely on local things. And it was good. But now it is not.

Here she overtly links climate change with cultural decline, using the English phrase "custom change" as her gloss for what I have been calling modernity the trickster ("I think the climate change is like the custom change. It's the same thing."). She makes clear the reasons for this analogy: both discourses posit decline ("It was good. But now it is not.") caused by Marshall Islanders ("Us people. Including Marshall Islanders") abandoning their culture ("local things") in favor of American culture ("we follow American culture") and relying on foreign goods ("[motor]boats and high buildings," "all these new things").

This overt, verbalized association is a rare instance. But instances of *implicit* association are rampant. I begin with the case study of an eroding coastal graveyard whose steady disappearance becomes, for concerned islanders, both a local impact of "climate change" and a sad illustration of abandoned custom.

CASE ONE: AN ERODING GRAVEYARD

A romantic western view of the South Seas might invest life before global warming—that atmospheric intrusion of industry into a primitive Eden—with a sense of sheltered tranquility and ecological security. But the Marshall Islands have always been vulnerable to oceanic and meteorological disaster (Bridges & McClatchey, 2009: 145; Erdland, 1961[1914]: 17–18; Kramer & Nevermann, 1938: 23–24). A 1905 typhoon obliterated three of Mili Atoll's islets (Spennemann, 1996), leaving nothing but a bare, submerged reef platform. A 1918 typhoon drowned 200 Majuro Atoll inhabitants and cut a stretch of land into several separate islands (Spennemann, 1996). Another storm in the mid-nineteenth century devastated Ujelang Atoll, leading a visiting westerner to assume that a volcanic eruption must have occurred (Spennemann & Marschner, 1994: 8). Early accounts of the Marshall Islands, written more than a century and a half before contemporary concerns about global warming-induced sea level rise, point to local fears of inundation that now seem remarkably prescient: if islanders were remiss in certain ritual duties, reports the naturalist Adelbert von Chamisso, they believed that "the sea would come over the island and all land would disappear. A well-known danger threatens all low islands from the sea, and religious belief often holds this rod above the people" (Chamisso, 1986[1821]: 278). The threat was perceived to be particularly grave in the windy, wavy *añōneañ* season. An islander told the visiting Russian explorer Otto von Kotzebue:

> The wind in the months of September and October, sometimes blows from the S.W., and not seldom rises into a furious hurricane, rooting up the cocoa and bread-fruit trees, desolating the islands on the western point of the group, which, he assured me, were sometimes swallowed up by the waves. (Kotzebue, 1821: 147–148)

So Marshall Islanders had reason, centuries before the phrase *global warming* was coined, to develop protection strategies against the threatening ocean. They employed safeguards ranging from settling on the most secure islands (Spennemann, 1996), reclaiming land (Bridges & McClatchey, 2009: 145), forecasting weather (*katu*), and using coastal trees to block wind and salt spray and to reduce erosion (see Chapter 5). They also had supernatural means of redress. Magic (*anijnij*) is a long-standing and still vigorous practice, used for everything from ensuring romantic success to killing an enemy to calming the seas. Adelbert von Chamisso, after noting local fears that the archipelago could sink beneath the waves, reports that "conjuring helps against this . . . Kadu [a local islander] saw the sea rise to the feet of the coconut trees, but it was abjured in time and returned to its borders. He named two men and a woman for us who understand this conjuring" (1986[1821]: 278). A century later, the German missionary and ethnographer August Erdland described an elaborate magical formula for protecting islands from high tides and storm surges (1961[1914]: 320–321). A few contemporary Marshall Islanders believe that protective magic of this sort is the reason why the Marshall Islands still exist; if not for these powers, the islands would have been destroyed by the sea many centuries ago. Reportedly, some still perform these spells. The magical preparations are called *jabwi* (cognate with *tabu* in Polynesian languages)—one magically treats a bottle and then buries it in the ground at the place one wishes to protect. It is said, "Jabwi so that waves don't come" (*Jabwi bwe en jab itok ṇo*). Some houses and a pandanus grove on the coast of Majuro's urban center are said to be protected in this way and to have been unscathed by the December 2008 flood as a result.[2]

This section concerns a particular Marshallese clan (*jowi*) associated with this kind of magic. Every Marshall Islander belongs to a *jowi*, of which there are at least thirty in Marshallese society (Kabua, 1993: 7). The basic rule for clan membership is matrilineal, but in practice such membership is more complex and bilineal. An individual is only truly of the mother's clan but also weakly of the father's clan; the clan's essence can pass through a woman without weakening, but it passes through a man only in diluted form. Thus one is strongly of one's mother but only weakly of one's father, the same philosophy governing Marshallese land tenure and inheritance more generally. Fellow clansmen were traditionally considered allies, and clan exogamy is still enforced today. Each clan is also known for certain skills or personality traits possessed by its members (Carucci, 1993: 161–162; 1998: 16) or for former ritual obligations and prerogatives in pre-Christian Marshallese society. *Mokauliej* clanspeople, for instance, are known to be artistic and to have good handiwork. They are too humble to ask for food even when they are hungry. Traditionally they were able to break certain taboos, such as not eating on graveyards. The *Jibuklik* clan once had a special relationship with chiefs: they acted as servants to them, walked freely in their midst, and dug their graves when they died (Petrosian-Husa, 2004: 48). In former times clans had animal or plant totems and were ranked in relation to each other (Erdland, 1961[1914]: 116–117, 343–345), but these associations are now largely forgotten. The importance of clan membership in the Marshall Islands has decreased since the advent of Christianity. Nowadays not everyone knows the reputation of his or her *jowi* or even its name; indeed, one study concluded that one third of urban Marshall Islanders between the ages of ten and twenty-five were not familiar with the word *jowi* despite being native speakers of Marshallese (Walsh, 1999). Others remember their *jowi*, and clans remain important for determining eligible marriage partners (see Carucci, 1984: 152).

One clan with a strong reputation is the Ripako[3] of Jaluit Atoll. Now considered extinct, their reputation lives on among many Jaluit residents and those who claim Jaluit Atoll Ripako ancestry if not true clan membership.[4] *Ripako* means Shark People, and the association of the clanspeople with the country's most ferocious predator is matched by their reputation for supernatural potency. Though the Ripako were not the only ones able to magically influence the ocean, their power over the weather and the sea is considered to have been particularly prodigious. They could summon or banish storms, typhoons, wind, and waves. This included the ability to stop sea level rise: if waves were threatening to rise to the level of the island, the Shark People could calm them.

The Shark People are said by their Jaluit Atoll descendants to hail from Piñļap, a large (by local standards) islet in Jaluit Atoll. Many stories of their magical powers take place on this home island. The clan's best known tale is said to have occurred in historical time, in the seventeenth, eighteenth, or nineteenth centuries. An American or European sailing ship arrived on the uninhabited ocean beach of Piñļap Island. The visitors dropped anchor and came ashore. There they found a boy fishing alone. They abducted him, returned to their ship, and set sail in haste. Their choice of captive, however, was unwise, because the boy was the son of a Ripako chief. When the chief noticed the boy's absence several hours later, he was furious. The ship was now only a speck on the western horizon, but the chief was undeterred. He summoned a storm by blowing on a magical conch shell; a wind came from the west, powerful enough to snap coconut trees, and blew the ship back to Piñļap, wrecking it on the reef. All aboard were killed except the boy, who was reunited with his father. The ship's anchor is still visible on Piñļap's ocean side, along with chains and iron bars.

Everyone who tells the story claims that it is literally true. A few of the tellers say that the storm may have been a result of good luck rather than magical power, but no one disputes the veracity of the events themselves. This is in contrast to most Marshallese legends (*bwebwenato* or *inọñ*), which are often said to be "just stories" (*bwebwenato bajjek*). Marshallese legends usually take place in a distant, immemorial past, while the story of the kidnappers occurs in historical time. In addition, there is physical evidence of the shipwreck on Piñļap, and some informants can even trace their genealogical relation to the characters in the story. So the tale of the shipwrecked kidnappers is told not just for entertainment but as proof of the Ripako's former powers.

The Shark People had a special, probably totemic relationship with their animal namesake. Some say that it was forbidden (*mọ*) for Shark People to eat sharks, because to do so would be "like eating their own flesh and blood" (Petrosian-Husa, 2004: 47), depleting the clan's magical power. Some say that before Christianity the shark was like a god to the Shark People, and that the clan is descended from a shark ancestor. It is often said that Shark People did not fear sharks—that they could go as far as to feed one of the hungry predators out of their hands without being bitten.

Ripako clansmen of the past, and their descendants today, are said to be physically imposing, strong, and tall; in former times they may have been giants. (The ancestors in general, not just the Shark People, are often said to have been very tall.) Rekadu, the kidnapped boy in the story, is said to have grown into a strong and muscular man possessed of prodigious magical skills. Juda, his sister's grandson, is described as tall, big-boned, and heavy-set, with hands the size of catcher's mitts; he is said to have been a master of *Kōtaan Eṃṃaan Jibbukwi* ("between one hundred men"—a kind of Marshallese martial arts), able to single-handedly best a team of men in feats of strength, or to sail alone in a canoe without food or water.

The magical power of the Shark People was invested in a particular site: the Bōn, a graveyard on the northern shoreline of Piñļap Island, where the Ripako ancestors are said to be buried. Graveyards carry great cultural significance in the Marshall Islands. In pre-Christian times Marshallese buried only chiefs on land, while dead commoners were sent out to sea. Nowadays Marshall Islanders bury their dead in the Christian fashion, in gravesites around the house or in separate cemeteries. While it is not uncommon to see children nonchalantly sitting on these newer gravestones, ancient cemeteries cannot be so casually treated, being possessed of powerful taboos. Marshallese say, "Your head is your graveyard." As in many Pacific societies (Firth, 1963[1936]: 15), the head is considered the highest and therefore the most sacred part of the body. Marshall Islanders say that one should not touch another person's head because it is precious (*aorōk*) and that one should not walk at night on the side of the house where people's heads are lain down to sleep. Just as a graveyard is forbidden, so too is one's head. Thus the statement that one's head is one's graveyard points to both the sacredness of the head and to the sacredness of graveyards.

Sacred graveyards are called *mọ* (forbidden, taboo) or *mọṇmọṇ* (possessed of spiritual power; haunted). These sites should not be visited casually: depending on the graveyard and the informant, one will hear that no one should visit the graveyard, that only the kin of the deceased should visit, or that strangers must be escorted by such a relative. In addition, one should not shout or play there. Dire consequences will result from breaking these prohibitions—one will sicken and die, or a terrible storm will be created. Some locals dismiss these taboos as old superstitions or say that they were once in force but have now weakened. But many fervently trust the taboos surrounding graveyards. A man from Ujae Atoll warned me not to visit the graveyard of his kin on Ebeju Island, saying that a man who had broken this rule had a terrible vision of angry spirits cutting his body into pieces and putting those pieces into a food basket.

The Bōn fits these general remarks about old graveyards; predictable taboos are associated with the site. Most people say that one must be escorted to the graveyard by someone related to the Ripako clan and that one should not make excessive noise while visiting. Breaking these rules will result in a powerful storm that prevents one from leaving the island. Other informants say that these magical sanctions are no longer in force because of the loss of Ripako power.

The Bōn is both the source and an indicator of the Shark People's power. In some versions of the kidnapping legend, the infuriated Ripako clansman performs the storm-summoning magic at the Bōn. One man said that the graves were placed close to the shoreline because the Ripako power would help to stop typhoons that could damage the island. Many informants claim that the graves are exceptionally long: 8 to 11 feet. A favored explanation of this is the huge size of the Ripako themselves or of the Marshallese ancestors in general. An archaeological survey of the site does not confirm that the graves are unusually large (Deunert et al., 1999), but the more important point is that people are convinced they are, and the impressive size is yet another indicator of the Shark People's erstwhile power.

Erosion is now dramatically evident in the Bōn and along the surrounding northern lagoon shoreline of Piñļap Island. When I visited the site in August 2009, numerous coconut trees had fallen into the water around the Bōn. The graveyard directly abuts the shoreline; indeed, it was difficult to determine where the Bōn ended and the shore began, as the beach was actively invading the graveyard, reclaiming it piece by piece. Descendants of the Jaluit Atoll Ripako say that the land used to extend to a rocky outcrop in the lagoon, 10 feet from the current shoreline. They are unable to say exactly when the erosion of the Bōn had begun, but most agree that it started in the recent past. Various artifacts have appeared from the Bōn as it erodes, including a conch horn said to have been blown by the chief to summon waves to bring back his kidnapped son, and human bones, said to be very large. On the ocean side of the island, a 10-foot-long iron anchor lies rusting on the shallow reef. All agree that this is a remnant of the ship destroyed in the legend of the kidnapped child, and it serves as another indicator of the Shark People's former might.

For locals concerned with the site, the erosion of the Shark People's source of power is no great surprise. Like so many things in the Marshall Islands, this clan and their powers are said to be disappearing or already gone. People point to many symptoms and signs of this decay. The clan's home island of Piñļap has been all but abandoned; it was once home to perhaps one hundred people, but now on this sizable piece of land there is only a single house compound inhabited intermittently for harvesting copra. Part of a more general decline in magical efficacy, the Bōn is losing its power and its taboo status; many say that the old supernatural sanctions against improper visitation are no longer in force because the powers of the clan, or the spirits of their ancestors, have vanished. One informant told me that the young people have become more assertive than their elders: when there was discussion, a few years previous, of how to save the Bōn from erosion, the young people were in charge, suggesting modern means of protection such as seawalls but never Ripako magic. Clans in general have declined greatly in importance, and some Marshall Islanders say that losing knowledge of one's clan will make one lose knowledge of Marshallese custom in general (RMI Government, 1996: 7) and even to forget the most fundamental part of Marshallese custom, which is to take care of one another (*lale doon*).

Another explanation given for the decline of the Shark People is that they failed to pass on their knowledge and power. To understand this extinction, recall that in Marshallese society one is partially of one's father's clan but only truly of one's mother's clan; a clan's essence can pass through a woman without weakening, but it passes through a man only in diluted form. The Shark People ended not with a bang but a whimper. The last real Ripako clansman on Jaluit Atoll was a man named Lobokto, born perhaps in the eighteenth century. He married Lijade, from the Rimae clan, meaning that their children would be truly of the Rimae clan, not the Ripako clan. Lijade and Lobokto had three sons but only one daughter, named Benam. (Some tellers of the kidnapping story say that the stolen boy was Rekadu, one of Lijade and Lobokto's sons, and the boy's magical rescuer was his father Lobokto.) The four children did not possess full Ripako powers because it was their father, not their mother, who was a Ripako; nonetheless, they did retain some of the power. The one daughter, Benam, was the best candidate to pass on these remaining Ripako powers, since she was female. She had two sons and two daughters, who, like her, had partial Ripako powers. These two daughters would have passed on the remaining Ripako powers intact, but one of those daughters, Neito, failed to produce any children, and the other, Arenbok, had only one child, a daughter named Maññe. Maññe was the last one with the Ripako powers, which were still strong enough in her to be able to stop a typhoon in its tracks. But she had no children, so the powers vanished with her death.

The Ripako essence had been thinned each time it passed through the bottleneck of a male ancestor—in the words of one informant, "like Kool-Aid that you keep adding water to"—until it disappeared altogether. Insufficiently zealous in producing heirs, the Ripako had allowed their powers to be lost. As one man said, "Now the Ripako clan is no more. It is no more. It's over. It's used up. It's gone" ("*Kiiō emaat Ripako. Emaat. Ejemḷọk. Emaat. Ejako*"). The other lines of descent in the genealogy lead to several of my informants, such as Alden Jacklick, Hilda Heine, and Carl Heine, who consider themselves related to the Shark People but would never claim to have retained any of the ancestral powers. At the site of the Bōn, when I asked one of my informants what would become of the graveyard in the future, he answered:

> It will be gone, all of it. There is no power left in the Bōn. In the past, you wouldn't be able to do what we are doing now, standing here. Before, one should not come here or make noise here. It would cause a storm. But now you can, because the power is gone. Look at these graves. They are very long—eight or nine feet. That is how tall people were in the past. Now they are short, and weak.

The demise of the Shark People is, for Jaluit people, part and parcel with the demise of custom (*ṃanit*) in general. The fate of the Shark People is a particularly distressing example, because the Ripako represent former Marshallese vitality in all of its glory: they commanded both the fiercest predator and the most powerful force of nature. There was once a way that Marshallese people believed they could protect themselves from sea level rise, and that way was the magic of the Shark People. But now the clansmen are extinct, their vitality dissipated, their island overgrown and all but deserted, while the graves of their ancestors and the place of their magical power are eaten away by the same rising waters that they once believed they were able to stop.

Locals provide several different responses to the question of why the Bōn is eroding. The first explanation for the erosion is that it has been caused by local development projects: on Jabwor (Jaluit Atoll's most developed and densely populated community), lagoon dredging and the construction of a new dock; and on another islet, the artificial closing of a channel linking a pond to the sea. These seem to have triggered erosion: not only Piñḷap, but many islands in Jaluit Atoll are said to be eroding. The second explanation for the eroding Bōn is sea level rise from global climate change, drawing on the scientific concept with which most of these Jaluit Atoll Ripako descendants are familiar. The third explanation is the loss of Ripako power. As the Shark People's power over the ocean wanes, their graveyard is eaten by the sea, and as the graveyard is eaten, the Shark People's power over the ocean diminishes: so the erosion of the Bōn is both a cause and an effect of diminished vitality, creating a vicious cycle. The fourth explanation is a lack of adherence to tradition. Women identifying with the Arno Atoll Ripako clan theorized that the Bōn was eroding because Shark People from Jaluit Atoll had broken the traditional taboo against eating shark. A man from Jaluit Atoll blamed the erosion of the Bōn more generally on the modern abandonment of tradition: "Everything is changing . . . [The erosion of the Bōn is] because of a lack of knowledge from our ancestors. People now are really influenced by western styles. They don't care about the old styles. They like things like bikinis. After World War II, Marshallese techniques disappeared. 1950s, 1960s—gone. Now they say 'walk' instead of *etetal.*"

These four explanations for the eroding Bōn—development, climate change, weakened magic, and cultural decline—are more alike than different. Climate change is akin to development because both are seen as part of modernity; in Marshallese discourse, both are the result of following *ṃantin pālle*, the American way, and abandoning *ṃantin ṃajeḷ*, the Marshallese way. Loss of the Shark People's power is one example of many. Thus, ultimately, all four explanations for the erosion of the Bōn can be understood as laments about modernity the trickster and cultural decline. The Jaluit Atoll Ripako descendants have thus found it effortless and intuitive to make this "impact of climate change" into yet another subnarrative, yet another example of the metanarrative of pervasive decline. The next two cases show that other Marshall Islanders are making the same connection, rendering the scientific idea of climate change easy to trust.

CASE TWO: COMMUNICATING CLIMATE CHANGE

"If they haven't heard about it yet, don't tell them." Such was the advice of a well-intentioned westerner before I left for the field to study Marshallese views of sea level rise and climate change. But the ignorance-is-bliss school of climate change education has no adherents in the Marshall Islands. Every local I asked was adamant that their compatriots should be educated about the threat of global climate change. One well-educated informant called the attitude of the aforementioned westerner "condescending." Eldon Note, the mayor of Kili, Bikini, and Ejit, was unequivocal when I interviewed him in 2007:

> I think [climate change] should be taught from elementary school up [in the Marshall Islands] . . . And I think it should be one of the [issues] discussed among all local governments and national government. I think it needs to be there as soon as possible. I think it should be yesterday or the day before yesterday.

As we saw in Chapter 2, several organizations in the Marshall Islands have taken up the call. Occasionally the communicators are foreign: in Chapter 5 I will narrate a notably unsuccessful communication attempt by a visiting Tongan man from the Pacific Council of Churches. But most outreach events have been locally organized and well received. While the scientific message of climate change is ultimately foreign in origin, it has arrived to ordinary Marshall Islanders via local communicators in an appealingly premasticated form; a form both easy to conceive of and easy to trust, as we will see.

The Ministry of Education's climate change-themed Education Week in February 2009 was a pioneer of its time, delivered at the vanguard of the many outreach events of that year. Like so many local organizations, initiatives, and political campaigns, the event was lent an air of traditionalism by christening it with a word from *kajin etto* (the old Marshallese language)—in this case *dienbwijrok*, a rare term referring to the last meal eaten together before a disaster (Abo et al., 1976: 55). (The full name of the Education Week was *Dienbwijrok: Jeḷā bwe jen pojak* ["*Dienbwijrok*: Know so that we can prepare"].) The events began with a climate change forum for students. Jorelik Tibon, the deputy chief secretary in the office of the president, defined the term "climate change" and told the youth in unequivocal terms that the threat was real. He explained that planetary warming was melting ice at the North Pole, raising the level of the ocean, and that the Marshall Islands would be severely impacted. Reggie White, the meteorologist in charge at the Majuro Weather Service Office, showed a graph of rising sea levels in the Marshall Islands and explained how factories, cars, and other industrial devices issued from abroad were responsible for this upward trend. The students were then broken into groups and challenged to answer the question "What can we as a community or as individuals do to combat climate change?" All of the groups decided that Marshall Islanders should wean themselves off of these destructive foreign artifacts in favor of benign local equivalents (see Chapter 4). This theme continued at an energy fair under the slogan "Conserve Energy Now!" Students were given pamphlets in Marshallese listing ways to reduce energy consumption, partly for the purposes of combating climate change. A tree-planting activity at Marshall Islands High School gave students hands-on experience in a traditional adaptation measure: the maintenance of a wind-breaking, salt spray-blocking, soil-binding shoreline forest (see Chapter 5). Students planted trees and bushes along the shoreline of the school, using their native names—*utilomar* (*Guettardia speciosa*), *wōp* (*Barringtonia asiatica*), *nen* (*Morinda citrifolia*), *kieb* (*Crinum asiaticum*), *arm̧we* (*Pipturus argenteus*), and others—emphasizing that these were native species with a long history and that their contemporary absence from the Majuro shoreline was a result of the wider disappearance of local things.

Education Week organizers also arranged a staged debate, on the subject of climate change migration, between Majuro's two tertiary institutions, the College of the Marshall Islands (CMI) and the University of the South Pacific (USP). As young people who might one day take part in this exodus, the issue was close to their hearts. The proposition—"Preparing Marshallese Citizens to Immigrate is the Necessary Response to Climate Change"—was to be supported by the CMI team and opposed by the USP team. Over 100 citizens made up the audience. CMI began with a pragmatic argument: "What will you do if your island is under water? Not only that but your family, your dog, your cat, your pigs, your chickens. We are living on low-lying atolls and islands, and the islands are vulnerable . . . Climate change is inevitable." USP countered: "We are Marshallese, we are proud to be here, and we will remain here for the rest of our lives"—prompting cheers from the audience. CMI rebutted:

> Why don't you go back to your house, take a yardstick—if you were here ten years ago and you claim yourself Marshallese, then you should really see the change . . . Do you guys know about Tuvalu? It's going under water. And do you know where Tuvalu is? It's in the Pacific Islands . . . It's too late to reverse it . . . The normal rate of rise of sea level is about 5.6 millimeters a year . . . It jumped to 10 millimeters . . . In one hundred years, it possibly might jump from eight to ten feet. Do you think we can survive that? I don't think so . . . It's also causing . . . gigantic floods, one whole city being under water. And not only that, there's also droughts. Every three to five years we get droughts, and it's getting worse. If we don't die by drowning, we die by dehydration . . . If we don't migrate, either we have gills and we can swim under water or breathe under water.

The audience laughed and cheered. The USP speaker, this time a young woman, countered by pointing to government efforts at climate change mitigation and weather forecasting, and then continued, "If we were to migrate, we would lose our land, our culture, our identity"—there was some clapping from the spectators—"our tradition, and especially our language. We would be aliens and treated like Indians . . . we would be considered second-class citizens . . . In our own country we are first-class citizens!" Spectators cheered heartily and a man in the audience said *Eañ, eañ* ("Yes, yes"). She continued,

> We might as well continue to live on the islands and use our health service facilities, our local medicines. Here in the Marshall Islands local medicine is free . . . It's all around . . . And you can just say *koṃṃooltata* ["thank you"] and present a token of appreciation and not pay thousands of dollars. Thirdly, we should play a part to slow down and stop global warming, altogether. We should cut down on electricity consumption by turning off air conditioners and enjoy our fresh air. We should build up seawalls to protect the land. We should not drive around unnecessarily.

At this point the CMI speaker's voice began to waver and she appeared tearful. "We should plant more trees! And we should create awareness of climate change!" CMI responded, "Do you have full trust in the Marshallese people? When they emigrate, they could still survive. Your culture still stands no matter what." This earned a few loud cheers and claps. "Do you trust the Marshallese people, even if they go to other countries, they could still survive because they have rich culture and abilities?" The USP team managed to twist this question to its favor: "We trust Marshallese people . . . By working together, we can pull ourselves from this disaster." CMI offered closing remarks: "We still believe that emigration is the necessary response to climate change . . . If you don't live, then you can't keep your culture. If you're going to die, where is that culture going to go to?"

In the end it was the antimigration side that was declared the winner. The debate's four judges felt that both sides had argued well, so they asked one of the judges, the climate change activist Mark Stege, to break the tie. As he told me later, given the strong performance of both sides, he had to decide according to the prevailing sentiment among Marshall Islanders on this issue. That prevailing sentiment was, as I will discuss in Chapter 5, against the idea of relocation.

In so many ways Education Week 2009 had borrowed from the conceptual structure and the values of the "modernity the trickster" narrative. The cause of climate change was overreliance on foreign things; the consequence was not only physical extinction but cultural death. Even the debate revealed that consensus: both sides presented themselves as being the voice of reason in the defense of threatened *m̧anit*—the promigration side by saying that tradition could not survive if the people drowned and the antimigration side by saying that they were protecting Marshallese culture from a culturally devastating move. The value of Marshallese tradition was constantly affirmed: the promigration side argued that the richness of Marshallese tradition would allow it to survive abroad and stated proudly that they trusted Marshall Islanders to maintain their culture despite environmental exodus; meanwhile the antimigration side argued that the richness of Marshallese tradition, its reliance on good native things ("our local medicines") and a convivial, unmoneyed lifestyle ("local medicine is free . . . It's all around . . . And you can just say *kom̧m̧ooltata* [thank you] and present a token of appreciation and not pay thousands of dollars") were the reason it should never be surrendered to evacuation. Never were foreign countries blamed: when it was asserted that climate change could be stopped, it was always actions by Marshallese citizens or by the government that were cited, and mitigation was equated, as usual, with rejecting foreign artifacts ("We should cut down on electricity consumption by turning off air conditioners and enjoy our fresh air . . . We should not drive around unnecessarily"). The two sides differed only in their level of optimism about *m̧anit*'s fortitude.

WUTMI was the other major player in 2009's spate of climate change outreach activities. The organization's climate change forum in April 2009, attended by about seventy women, began with a presentation by Yumi Crisostomo, then director of the government's Office of Environmental Planning and Policy Coordination (OEPPC). She stated that climate change posed a severe threat because Marshall Islanders live so close to the ocean. *Mantin etto* ("the old way"), she said, was to *mour jān lo̧jet* ("live from the sea"). She summarized the impacts expected in the next 50 or 100 years: a temperature rise of 4 degrees Fahrenheit, coral bleaching, increased droughts and extreme weather, and several feet of sea level rise. As a result, she said, *m̧anit ko ad* ("our customs"), could disappear. If there is no Marshall Islands, she asked, where will we take our culture? While other low-lying island nations, she said, were searching for land abroad to resettle, the Marshallese government considers such a response unacceptable. Industrialized nations were largely responsible for the problem, she stated, and should therefore provide funding for adaptation, just as they paid compensation for nuclear harms. Crisostomo summarized the Marshallese government's plans to prepare for and combat climate change, including energy policies, environmental monitoring, conservation activities, and campaigns abroad. She also suggested measures that ordinary Marshall Islanders could adopt, such as conserving energy and water, planting trees, avoiding the use of Styrofoam, and turning off air conditioners.

Moriana Phillip, from the Environmental Protection Agency (EPA), was next to speak. She said that women were especially vulnerable to climate change and its impacts on water resources owing to their customary roles, in which they use water for cooking, cleaning, washing children, and otherwise taking care of the family. Climate change, she said, could destroy the land of *jibbwid im jim̧m̧aad* (literally "other grandmothers and grandfathers," figuratively "our ancestors") and doom *m̧antin m̧ajeļ*. As just one example of this, particular plants that grow near the shoreline, necessary for traditional medicine given to postpartum women, could be wiped out by coastal erosion. While Kiribati and Tuvalu had expressed a willingness to abandon their islands and move to Australia or New Zealand in fifty years, Marshall Islanders reject this idea, she said. Individual actions that could combat climate change and reduce its impacts included recycling water containers, building seawalls, conserving energy, safeguarding marine resources, and working with industrialized nations to reduce greenhouse gas emissions. Terry Keju, the EPA's deputy general manager, emphasized coral bleaching and sea level rise as climate change impacts that threatened land- and sea-based subsistence. Tradition depended on the land and could not be simply exported to a place like Arkansas.

In July of that year, WUTMI held another climate change forum. Marie Maddison introduced the topic of the conference: "What can we do about climate change [*oktak in mejatoto*], changes in weather [*oktak in lañ*], and caring for our surroundings [*kōjparok peļaak ko peļaakid*]?" Angeline Heine, the national energy planner at the Ministry of Resources and Development, explained the difference between renewable and nonrenewable energy, saying that the Marshall Islands uses large amounts of nonrenewable energy imported from *laḷ ko rōḷḷap*, "the big countries," the industrial nations. Michael Honeth, a non-Marshallese working for the EPA, speaking through a translator, stated that the climate change one sees today comes from human activities. Climate change impacts, he said, are also exacerbated by local actions such as dredging, pollution of the reef, deforestation, and overfishing. Marshall Islanders, too, added greenhouse gases to the atmosphere. (The translator added: "Our lives and our tradition will be damaged.") Although there was not much that such a small country could do, he said, Marshall Islanders could work with other countries, conserve their own resources, curtail destructive development practices, limit dredging, and encourage each other not to throw garbage on the reef; but such actions, he added, could be difficult because Marshall Islanders, like people everywhere, were tempted by modern conveniences such as cars and air conditioning.

I then presented in Marshallese, explicitly suggesting a traditionalist interpretation of climate change. Acutely aware of the fact that I was not Marshallese and wary of the potential antitraditional air that this fact could lend to my speech, I utilized the "oracle effect" (Bourdieu, 1994: 211): I presented my comments not as personal opinions but as the collected, synthesized consensus of the many Marshall Islanders I had spoken to about climate change. The fact that WUTMI had invited me to speak and that I was speaking in Marshallese also contributed to my credibility. Marshall Islanders had told me, I said, that the loss of the islands would spell the demise of the culture, that monetary compensation for this loss would tragically reminiscent of the country's nuclear past, and that it was therefore imperative to prevent such a dreadful scenario. Changing the country's energy base represented only a tiny action in global terms, but it could nonetheless be symbolic of commitment and concern vis-à-vis climate change. Marshall Islanders, I argued, had dealt with sea level rise in the past, through their use of Ripako magic. If supernatural incantations were the old method of combating sea level rise, a newer but still "traditional" method could be to adopt renewable energy and revitalize coastline forests. Technologies like solar power, I said, while not part of the old ways, were still "traditional" in that they relied on local resources—*ejjeḷọk wōṇāān*, they are free of charge, like all things in *ṃantin ṃajeḷ*—and they produce zero carbon emissions. In this way, renewable energy was consonant with *ṃanit*, and caring about tradition meant caring about climate change.

Daisy Alik-Momotaro concluded the forum by reiterating that it is "we" human beings who are the cause of climatic damages; the Bible, she said, states that we ought to be good stewards of the earth.

In August, WUTMI hosted another outreach event, its executive board meeting (despite the name, this was a workshop attended by many WUTMI members besides the board). Unfolding over the course of six days, the meeting included many presentations on issues other than climate change; I will focus on the parts of the conference most directly relevant to climate change. At the opening ceremony, respected traditional leaders, including Lerooj (female chief) Neimat Reimers and Iroojļapļap (paramount chief) Mike Kabua, were in attendance, sanctioning the event with chiefly authority. The national anthem was sung:

> I love the islands where I was born
> The surroundings, the paths, the gatherings
> I will never leave because it is my true home
> And my inheritance forever
> It is best that I die there
>
> *Ij iọkwe ļọk aelōñ ko ijo iaar ļotak ie*
> *Meļan ko ie, im iaļ ko ie, im iaieo ko ie*
> *Ijāmin ilọk jān e bwe ijo jikū eṃool*
> *Im aō ļāṃoran indeeo*
> *Eṃṃanļọk ñe inaaj mej ie*

Mona Levy-Strauss offered some opening remarks, weaving together climate change with other signs of decline:

> Things are not good for us Marshallese these days. There is more disease—diabetes and cancer—and we don't teach our children *ṃanit*. There is climate change now in our surroundings, and increased warming in the world. It is not far away now! WUTMI is doing its part about climate change by hosting meetings. Our islands are low, and damages can come quickly to us from sea level rise. We Marshallese women need to come together and help each other.

Another speaker in the opening ceremony, the Honorable Matthew, stated:

> We must remain Marshall Islanders. Even if we go to America or Japan for school, when we come back we must still be Marshallese. We need to hold on to our *ṃanit*, because it's a heritage from our ancestors, and a gift from God . . . *Ṃanit* is important. We learn it from our mothers. But soon we'll be learning it from foreigners in schools! Marshallese life is a good life—respect is paramount.

A group of women performed a climate change-themed song composed by one of them, the woman whose statements I presented in the previous chapter in the section titled "Visions of Destruction." The song began with the proverb *Liṃaro pikpikūr kōļo eo*; this could be loosely translated as "Women, shake the spirit" and refers to the role of Marshallese women in lending morale and invigorating men and communities in their endeavors.

When “climate change” and “global warming” take effect, we will see changes in the world
The ocean and lagoon waters will heat up, and the food of the reef may disappear
If the warming is large, the wells will run dry, the plants of these islands will fade like the heat of the day fades in the evening

Ñe enaaj baj jejjet kūtien climate change global warming, jenaaj lo oktak ko an laļ in
Lọjet eo ilik im iar in aelōñ kein renaaj okmāāṇāṇ, im ṃōñā in ioon pedped remaroñ jako
Eļaññe enaaj ļap okmāāṇāṇ, aebōj laļ renaaj eļļaļļaļ, keinikkan ko i aelōñ kein renaaj aemedļọk

The following day saw several presentations on climate change. I gave a speech in Marshallese much like the one I had given in July. Yumi Crisostomo explained the causation of climate change: the atmosphere was like a blanket that was smothering the earth. She assured the audience that climate change was definitely true, that 95% of scientists agreed on this. Even if there was no hope and sea level rise would destroy the Marshall Islands, people needed to prepare so that they could somehow stay in the country. Although the big countries were the true culprits in this tragedy, Marshall Islanders could try to adapt. They could also pray to God. An EPA official reiterated Yumi’s statement about blame: “We have not made climate change happen. It is the big countries [*laļ ko rōļļap*], with their industry, that are doing it.”

Two days later, Angeline Heine’s presentation on “Energy and Gender” emphasized the difference between renewable and nonrenewable energy sources. The difference, in her account, was not merely that the former lasts forever while the latter does not, but also that the former comes from resources within the country while the latter comes from resources outside the country. “In the Marshall Islands we take a lot of things from outside the country,” she said. “90% of our energy comes from oil . . . ‘Climate change’ means *oktak in mejatoto*. It comes from using oil. This is making a hole in the atmosphere.” *Renewable* was thus equated with *local*, itself nearly synonymous with *traditional*; therefore even modern technologies such as solar power could be seen as authentic to *ṃanit*. Heine provided suggestions for conserving energy. Renewable energy sources like wind, water, ocean, and sunlight are never exhausted (*rejaje maat*), and do not damage (*kọkurre*) the *mejatoto* or waste (*kọkurre*) money. She passed out pamphlets in Marshallese describing twelve methods of reducing electricity usage.

All of these events, organized by WUTMI and Ministry of Education, were well received. When I talked to audience members during, immediately after, or months after the events, they always described them in positive terms; they were happy to speak at length about climate change (as was evident in many of the statements in Chapter 2) and almost always professed belief in it. For instance, after the July WUTMI forum, two *lerooj* (female chiefs) expressed their concern about climate change, said they wished to learn more, praised the forum for educating citizens about the threat, and said they would give serious thought to the role traditional leaders ought to play in the country's response.

I suggest that this positive reaction to a potentially repulsive idea stems primarily from WUTMI's and the Ministry of Education's framing of climate change in terms of modernity the trickster and pervasive decline. Climate change was presented as another on an already long list of threats to *m̧anit*. The abandonment of the archipelago was spoken of as an unacceptable resignation to cultural decline. The causes of climate change were identified as foreign artifacts, on which Marshall Islanders rely too much. There were, to be sure, a few discordant notes. Yumi Crisostomo from the OEPPC put primary blame for climate change on foreigners, in contrast to the usual in-group blame for cultural decline; I will have more to say about this differing governmental perspective in Chapter 4. Still, premises and concerns consonant with traditionalist decline dominated throughout. There was also, as we have seen, some use of Christian rhetoric, but it was less salient than traditionalist rhetoric, following this country's usual pattern, which I mentioned previously.

This locally appropriate rhetoric, I believe, did much to bring climate change to wide public attention and acceptance in 2009. The increasingly popular framing of climate change as cultural decline may also help to explain why Marshall Islanders seemed significantly more at ease with the subject of global warming, more active in responding to it, and less likely to deny its reality in 2009 than in 2007 (see Rudiak-Gould, 2009a, for a summary of my findings from 2007); by 2009 the society had found a way to make conceptual peace with global warming; its popularity as a focus of conversation and action increased as a result.

CASE THREE: THE (MIS)TRANSLATION OF CLIMATE

A voluminous body of work shows that the concept of "nature" as a realm distinct from and opposed to human social life is a central ontological, epistemological, and ethical tenet of modern western society (Beck, 1992; Chakrabarty, 2009; Escobar, 1999; Ingold, 2006[1989]; Latour, 2004), radically different from the more holistic views of many indigenous and nonwestern societies (see, for instance, Cruikshank, 2005; Gold, 1998; Huber & Pedersen, 1997; Reichel-Dolmatoff, 1976). Climate change, with human-nature entanglement as its central premise, would seem to put the final nail in the coffin for that ontology (Chakrabarty, 2009; Hulme, 2010b; McKibben, 2006[1989]), heralding a new "postnature" zeitgeist (Beck, 1992; Giddens, 1999) where nature as a pure and unhumanized realm has been lost. But evidently even anthropogenic climate change has not changed western minds: the popular characterization of climate change as an "environmental" issue, our persistent eagerness to pigeonhole Hurricane Katrina as either "natural" or "human-made" (not allowing it to be, instead, a human-natural hybrid [Hulme, 2010b]), shows that even in the Anthropocene we are still thinking in terms of nature and culture (see Uggla, 2010).

Other societies encounter global warming without this ontological baggage. The Marshallese language has a word that can be glossed as "culture" (*ṃanit*, also translated as "tradition," "manner," or "manners"), but it has no word for "nature." Two words, *peḷaak* and *meḷan*, are now being used by Marshallese activists to refer to what we might call the "environment." But the words simply mean "surroundings" or "surrounding area," without a deeper symbolic dichotomy between the human and the natural. *Ipeḷaakin* (in the *peḷaak* of) simply means "around." In no sense is the Marshallese landscape considered "natural," since it is everywhere named, owned, and storied, and nearly everywhere cultivated (see Chapter 5).

This absence of a nature-culture dichotomy comes out in Marshallese discourses of decline. Up until now in this book I have (quite modernistically) divided Marshallese narratives of change into narratives of *social* change (Chapter 1) and narratives of *environmental* change (Chapter 2). But no such strict separation is evident in the statements of Marshall Islanders themselves. The disturbing trends that locals point to often blur the line between natural and cultural. Physical and supernatural vitality have waned. People are said to be physically smaller than they were in the past: the ancestors were giants, 7 or 8 or even 10 feet tall. Islanders are increasingly wracked by disease; this is blamed on radiation, on unhealthy imported food, or on the influx of foreigners who bring with them previously unknown maladies. As a result people live shorter lives; people used to live 100 years or more, it is said. In addition people these days are lazy, and their magic lacks the potency of yesteryear: once it was possible for a Marshallese magician to merely look at a coconut tree and say, "I admire those coconuts," and they would fall, or he might make someone's leg break with a word; but these skills have now vanished, locals say. Time itself is perturbed: "The earth is faster now" (Krupnik & Jolly, 2002) is for Marshall Islanders a literal statement: "The days are faster (*eṃōkajḷọk raan*), they used to be slow. It's noon now but very soon it will be night." "Time is really fast now. You chat and you don't know but it's already midnight!" "It used to be that you could sleep a long time, wake up, keep sleeping, sleep till you're sick of sleeping, and then get up, and it's still dawn. Now you wake up and it's already late morning. The daytime comes quickly these days."

Loss is everywhere and in everything. On Jaluit Atoll, it is said that the rats of Arbwe islet used to come and "greet" people as they arrived on the beach; but they no longer do this, and no one knows why. On Namdrik

Atoll, there was once a celebrated flower, a special variety of the *kajdo* flower found only there. But, my Namdrikese informant told me with sadness, he had returned in 2003 to find it inexplicably gone. On Ujae Atoll, a kind of coral called *luo* was once prized for jewelry making, but recently, when some local men attempted to find it in its expected location near Ebeju Island, it had vanished. On Mejit Island, an old woman told me that in the past, one could shout on the north half of the island and hear an echo; now that echo has disappeared. "Everything is gone," the woman said. The biblical end-time may well be at hand. The signs of apocalypse are already obvious: foreign wars, family squabbles, selfishness, illness, and strange perturbations in the cosmos. The Iraq War, swine flu, the 2009 solar eclipse, local land disputes, a hermaphroditic pig—all are disturbing intimations that the end of the world may be near.

Not surprising, then, that my questions about "environmental" change were often answered in broader terms by my informants. Interviewing an elderly man on Jaluit Atoll in August 2009 in the Marshallese language, I repeatedly broached what I considered "environmental" matters, only for him to respond with "sociocultural" laments:

PRG: Has the climate [*mejatoto*] changed since the past, or has it stayed the same?

X: It has changed. These days belong to Americans. Everything is different . . . People nowadays don't know the ancient way and the ancient language. They don't know Marshallese tradition. Marshallese tradition is on the brink of disappearing because of money. Money damages tradition. We don't feed people. We follow the American way now . . . We eat lots of [imported] rice now, and less Marshallese food. People don't know how to eat coconut seedlings or breadfruit. They only want to eat rice . . . There's no Marshallese food on this island now.

PRG: Has the ocean changed since the past, or has it stayed the same?

X: It has changed. And why is that? Because people seek money, so they hunt lots of clams, so there are fewer of them nowadays. Nowadays you use motorboats instead of canoes, and the price of gas is high. If you don't have a canoe, you suffer because you have to pay a lot for gas.

PRG: Has the weather [*lañ*] changed since the past or stayed the same?

X: The *mejatoto* [climate, air, atmosphere, cosmos] is different from the past. They used to perform *katu* [traditional weather forecasting]. If they said next month it will rain, they were right—it would rain. Now they don't know how. Only a few know, but they don't teach any children . . . No one is interested in learning traditional weather forecasting. They only want to learn American-style navigation.

Later in the interview I discovered that the man did indeed have opinions on what westerners would call "environmental" change and was fully aware of the scientific concept of global warming. So his "sociocultural" answers to my "environmental" questions did not stem from a lack of interest in "the environment," but from a different and more holistic ontology.

Global warming, too, is understood in this broader light, to the extent that sometimes it is difficult to tell if Marshall Islanders are talking about climate change, cultural change, or both at once. One eventually abandons the effort to differentiate between the two, as locals are not very interested in making the distinction. In an interview I conducted in 2009, a female government employee in Majuro who had attended the WUTMI climate change forum in April 2009 began with a litany of purely "environmental" observations:

X: We can see climate change with our own eyes. It's hotter—you go outside and it's hot. The soil is dry and salty from ocean water. Trees and plants are dying from the salt. Wells are running dry. This is all because of climate change. This is in the outer islands . . . In Majuro, there are problems with water supply. There's not enough water . . . Families see that their water catchments are empty . . . Plants are dying in Majuro because of the sunny weather.

PRG: What about in the past?

X: It used to rain at least once a week. The plants grew well. Nowadays it's really hot.

Then she segued seamlessly into a discussion of "sociocultural" changes, then back into "environmental" changes, and back once again, until the distinction between the two had evaporated:

X: One reason for the changes is that families don't have enough money. There are more people in each family. People live by money (*mour kōn m̧ani*). In a family there are only two people working but many in the family. So there's not enough money. People get sick from not having enough water and food. Also, the islands are getting smaller—you see them getting smaller and smaller . . . It started around 1985. When I came here from the outer islands, I saw the end of Laura and it was very long. Now it's not. I met a pastor from Mejit [Island] who says that coconuts and breadfruit there on Mejit used to be this big [gestures very large] but are now this big [gestures much smaller]. And the water catchments there are running empty. The breadfruit trees have no fruit. We're sure that this is because of climate change. The soil is bad too, from climate change.

PRG: Why is climate change happening?

X: Well, I believe in what the Bible says. There will be a time of big wars, sickness, the rich getting richer and poor getting poorer. People will go hungry. I read this in the Bible. So maybe it is the time for this. So we're scared but we know we cannot do anything about it.

PRG: So you're talking about the end-time?

X: It's coming closer. When we observe people nowadays, we see that they hate each other. That's not like before. We used to love each other, go together with each other.

PRG: Why has this change happened?

X: There's not enough food and not enough money to provide, and not enough space for everyone to live. And there's inflation, also, and too many people. The salaries are staying the same, but the prices of everything else are going up. [Eventually] climate change will make people leave the country.

PRG: Will anyone stay in the Marshall Islands despite climate change?

X: Maybe a few. We love our islands. We don't like Chinese! It looks like the Chinese are going to take over the country.

In my Majuro survey, I asked respondents who had reported changes in the weather why these changes had occurred. Many answers were along these lines: "Because tradition has lessened." "Because we don't cherish [*kōjparok*] our tradition." "Because people's lives are different, so the *mejatoto* is different." "Because life is harder these days. Fathers and sons hate each other." "Because life and culture in this country have changed, because of the things that come from foreign countries. We watch movies and copy the violence we see on them." "From what I hear, it's because life has become harder." "We don't depend any more on what we should depend on. And many Marshall Islanders have left the country." Other answers to this survey question blamed local climatic changes on the same sorts of foreign imports, the material trappings of modernity, which are seen to cause cultural decline. "There are more chemicals now. Technology." "Because there are lots of goods [*m̧weiuk*] nowadays." "It's because of life changes. Population growth causes problems in the environment. Cars contribute to pollution. There is more waste." "Some islands are different now because construction has damaged them. And there are no trees now in Majuro." "Because there are more engines, vehicles, and so forth." "There are lots of houses now, so there are no trees." "It's from the chemical emissions. These things come from the Majuro power plant. Nowadays Marshall Islanders only rarely use *um̧* [traditional earth ovens]." When I asked survey respondents who had heard of the scientific concept of global warming to define the English term *climate change*, the answers were often broader than standard scientific definitions: "Changes in the *mejatoto* [climate]. All things that happen in a country." "Changes in the *mejatoto*, life, and all sorts of things." "Changes in the *mejatoto* and changes in people's lives." "Changes in the world." "Everything changing from the original way that it has always been."

Climate change thus comes to be seen as a cause, an effect, and an example of cultural decline. In the Marshall Islands, it could hardly be otherwise. Even the linguistic translation of "climate" enforces this view. (For a more detailed treatment of this topic, see Rudiak-Gould, 2012a.) Climate change communicators in the Marshall Islands most commonly use the English phrase *climate change* or the Marshallese phrases *oktak in mejatoto* ("change of the *mejatoto*") or *ukoktak in mejatoto* ("continuous change of the *mejatoto*") to refer to the issue. Those locals who know the English phrase *climate change* almost always define it, when asked, as *oktak in mejatoto* or *ukoktak in mejatoto*, so *mejatoto* is the word that has been widely adopted as the Marshallese translation of *climate*, and when scientifically untrained Marshall Islanders say the English word *climate,* it is *mejatoto* that they truly have in mind. But the fit between Marshallese *mejatoto* and English *climate* is imperfect. To be sure, *mejatoto* very often is taken to refer to meteorological conditions such as wind, temperature, and precipitation. But that hardly exhausts its semantic breadth. The Marshallese-English dictionary also lists "air," "atmosphere," and "space" among its meanings, but even this is too narrow. In practice *mejatoto* can refer to the cosmos in general, of which sociocultural, environmental, and cosmic processes are all unseparated parts. The acceleration of time, which many locals report, is considered a change in the *mejatoto*. So is the 2009 solar eclipse, and a whole host of laments like increasing disease, laziness, selfishness, and the general badness of life: all were given in response to my question as to how the *mejatoto* had changed.

The result is not merely that climate change is understood a particular way—as part and parcel of general decline—but also that it is made eminently believable, obvious, almost banal. If all sorts of things can count as a change in the *mejatoto*, then the world is positively overflowing with evidence of climate change's reality—a process that I call "promiscuous corroboration" (Rudiak-Gould, 2012a). Recall the eclipse observers at the beginning of this chapter, "the eager observers" and "the hyperbeliever" from Chapter 2: the field of "observation" as a source of climate change information and belief had become extremely wide, so that even an event in outer space, or an unusually *low* tide from a tsunami, could be taken as evidence for climate change. On Jaluit village, Jaluit Atoll, three weeks after the eclipse, a middle-aged woman told me that the *mejatoto* had certainly changed: not only had the air gotten hotter, the sea inched closer to land, and the coconut trees fallen on the shore but also time itself had sped up and the sun had gone dark. The same day, a young Jaluit man told me, "Time is really fast now. You chat and, before you realize it, it's already midnight . . . Maybe it's part of climate change . . . These days you work only for money. People rely less on Marshallese food now . . . I think that's why there are plenty of typhoons, earthquakes, and illnesses [in the world]—because of climate change." When I asked a Majuro man if he believed what scientists said about climate change (using the English phrase), he replied, "I think it may be true. Because I see that the *mejatoto* is not very good nowadays. Life is harder. Goods are expensive. The sun is stronger . . . And there are improper relations between kin."

There is a helpful comparison here to *baijin* (radiation) as a Marshallese explanatory framework. Recall from Chapter 1 that locals blame many kinds of ills on lingering radiation from US nuclear testing: not merely cancers and birth defects, but also (for example) shorter life spans, diabetes, hermaphroditic pigs, and the December 2008 flood. Although certainly far less salient than the abandonment of tradition, radiation functions in the same manner as a catch-all explanation for disturbing events. Now climate change is beginning to supplant radiation in this role. For decades (see Spennemann, 1993: 144), the decline of arrowroot—a former staple crop now harvested in only small quantities—was most often blamed on radiation. During my fieldwork in 2007, this was still the case. By 2009, some locals had started to point the finger at climate change instead. Whatever the true cause may be, there is much to be learned from this attribution. Climate change, it seems, is the new radiation: the new inscrutable force, existing everywhere and nowhere, visible only as "impacts" that even scientific experts are hard pressed to predict, able in theory to disturb or to harm nearly anything. To not be able to *see* the causation anywhere is to be able to *perceive* it everywhere. Doubly so in a society where nothing is natural because there is no "nature": everything can be human-caused. The changing *mejatoto* is a thing so general, so holistic, that a society already preoccupied with decline would be hard pressed to seriously question it.

CONCLUSION

As we saw at the beginning of this chapter, much of the literature on climate change communication is concerned with the barriers to belief and concern and the enormous challenge of convincing the public that the climate crisis deserves attention. But perhaps the difficulties appear so formidable only because it is a western audience that we have in mind. In the Marshall Islands, climate change is not such an ideological affront to the distinction between the natural and cultural (Chakrabarty, 2009), the inevitability of progress (Norgaard, 2002), the goodness of contemporary life (Feygina et al., 2010), or the virtue of individualism and industry (Kahan et al., 2007; McCright, 2011). Never having had faith in those postulates in the first place—indeed, placing their faith in precisely the opposite—Marshall Islanders encounter climate change as an ideological compliment, not an insult. Perhaps one could even go further: the idea of the Anthropocene is so intuitive to a society already convinced of human-made woe that even if no environmental changes were occurring, even if scientists had never brought news of global warming, Marshall Islanders would nonetheless report that the weather was worsening and worry that their islands might sink. Voltaire famously said that if God did not exist, it would be necessary to invent him. In the Marshall Islands, if climate change were not real, people would have had to invent it.

4 Seductive Modernity, In-Group Blame, and the Mitigation Movement

Along with the tropes of paradise lost, sinking civilizations, and canaries in the coal mine, it is *innocence* that most often characterizes outsiders' portraits of islanders on the frontlines of climate change (Barnett & Campbell, 2010). As the anthropologists Susan Crate and Mark Nuttall write,

> Climate change is the result of global processes that were neither caused nor can be mitigated by the inhabitants of the majority of climate-sensitive world regions now experiencing the most unprecedented change. . . . Indigenous peoples themselves may argue that, despite having contributed the least to greenhouse gas emissions, they are the ones most at risk from its consequences. (Crate & Nuttall, 2009c: 11–12)

The plight of small island states is often cited as the epitome of this disparity (Barker, 2008), and the Marshall Islands seems among the best examples of all. The crime is a mere 0.0003% of global greenhouse gas emissions in 2008 (United Nations, 2011); the punishment is nationwide uninhabitability. It would take quite some mental acrobatics to dismiss the cries of climate injustice emanating from many indigenous leaders around the globe (H. A. Smith, 2007).

An outside observer might thus expect particular responses to the question that was posed to several hundred Marshallese students in Majuro on February 25, 2009: "What can we as a community or as individuals do to combat climate change?" As attendees at a Ministry of Education-sponsored awareness-raising session, the students had just heard the unpleasant news—familiar to some, novel to others—that human use of technology was raising the oceans and threatening to inundate the country within these young citizens' lifetimes. Expected answers might have included: "We can do little or nothing, because we have not caused the problem"; "We can prepare for impacts, but attempting to prevent them is futile"; "We must shame foreign polluters into cleaning up their act." Quite the contrary, the answers were the following:

> Instead of using gas—as we all know we use gas around the Marshalls for our vehicles—we can use biofuel, a.k.a. coconut power. . . . Instead of boats or *tiṃa* [ships] to go to visit grandma in the outer islands, we can use *kōrkōr* [canoes], because we are sailors.
>
> Reduce the use of private transportation, which means use more public transportation. . . . Reduce fossil fuel emissions. Use alternate forms of energy, for instance solar, wind power. . . . Don't burn trash. Recycle it. . . . Turn off lights, air conditioning, fans. . . . Reduce your use of electricity as a whole.
>
> Use less CO_2. We use a lot of CO_2. For example, reduce the number of cars, use less air conditioning . . . and reduce the number of batteries. These things all use CO_2. Do not wait. Act as fast as you can to reduce climate change. . . . Hurry, hurry, hurry. Don't just sit down and be lazy.

The position was unanimous: Marshall Islanders' response to climate change should be to reduce their own carbon footprint. Never was there a mention of protest against outside offenders, such as northern governments, affluent First World citizens, or corporate elites. As I will explore in this chapter, this response is representative of a larger pattern in the Marshall Islands, and its reasons have nothing to do with ignorance, denial, pragmatism, or timidity but instead with local understandings of a seductive modernity.

WHO OR WHAT IS RESPONSIBLE?

The first question that must be asked about Marshallese perceptions of climate change responsibility is not which humans they blame but whether they blame humans at all. In the United States, the individuals we call climate change skeptics often accept that warming is occurring but question its anthropogenic origins. Concerned with the policy inertia that these disbelievers have helped to entrench, a number of scholars have sought to identify the root cause of anthropogenic skepticism (see for instance Brugger, 2010; McCright, 2011; Oreskes & Conway, 2010). One proposal is that human influence on the climate is not so much an ideologically inconvenient truth as a conceptually counterintuitive one (Chakrabarty, 2009; Jasanoff, 2010; Moser, 2010: 34). Geographer Simon Donner offers the most sweeping claim: "virtually all religions and cultures worldwide for thousands of years" (Donner, 2007: 235) regard the climate as too grand to be under human influence: it is "the domain of the gods," a force that only deities can manipulate. Hence the public's incredulity in the face of a scientific consensus.

But Donner's argument is unconvincing. The idea that spirits and deities control the weather is indeed cross-culturally widespread, but so is the notion that human behavior can enrage or pacify these agents, thus influencing the weather at second hand (Byg & Salick, 2009; Gold, 1998: 174; Huber & Pedersen, 1997; Strauss & Orlove, 2003: 3, 7).[1] In many cultures *direct* human influence on the weather is taken seriously too: people's moral behavior (Hitchcock, 2009; Hsu, 2000; Leduc, 2007), magical formulas, or the activities of powerful foreign groups are seen to shape the weather (Byg & Salick, 2009; Hitchcock, 2009: 258). Worldwide, most people have no trouble accepting that human actions are now altering the global climate: in the 2006 Nielsen Global Omnibus survey, the highest percentage of anthropogenic disbelievers was 16% (for American men)—perhaps a regrettably large number but still a small minority; and this is the *highest* percentage of skeptics found in a survey of many nations. So Donner gets it precisely backwards: the cultural oddity, the strange new idea, is not the scientific claim that humans influence the weather but the modernist notion of "blind, pitiless indifference" (Dawkins, 1995: 133)—an unresponsive cosmos that cares nothing for the good or bad deeds of humanity. The older (some might even say "original"), cross-culturally intuitive view is that the weather and other natural processes register our actions, rewarding and punishing us (Hulme, 2009: 13–14; Rayner, 2003: 278; Strauss & Orlove, 2003: 3–4, 7).

Marshallese culture is no exception in this regard. As we saw in the previous chapter's study of an eroding graveyard, many Marshall Islanders believe that human magicians can influence the weather and the ocean, or at least that they could do so in former times. The idea that human trespasses can be punished with environmental disaster is certainly a longstanding one in this society, judging from Chamisso's observation in 1817 that islanders considered devastating sea level rise to be a punishment for ritual error (Chamisso, 1986[1821]: 278; also see Carucci, 1997b: 19). As I explored in Chapter 2, many locals believe that Americans and scientists possess prodigious powers, which could easily include weather control.

So we would expect Marshall Islanders to find it entirely plausible that human beings are responsible for global warming; and indeed they do, as we will see. The only other locally viable attribution strategy would be to "blame" God. Interestingly, only a very small minority of locals subscribe to such a view. Although in Chapter 2 we saw the Revelation argument that climate change is part of the End Times, this viewpoint is rarely expressed in terms of God "causing" climate change to occur. Furthermore, almost no one—only a few individuals out of hundreds that I talked to—considered climate change a divine punishment for human sins. In the Majuro survey, when I asked "Why is the *mejatoto* changing?" of the 100 respondents who had reported it was changing, only one individual attributed the change (tentatively) to God's will. In responding to the question "Who is to blame for 'climate change'?" only 5% of

77 respondents "blamed" God. As indicated in Chapter 1, for Marshall Islanders it is tradition, not God, that is the usual "grand arbiter," the entity that, if disobeyed, will cause suffering.

Blaming climate change on "nature" is not intelligible in the Marshall Islands, for a few reasons. First of all, as we have seen, the Marshallese language has no synonym for *nature.* Well-educated Marshall Islanders may refer to nature when speaking in English, but only occasionally, and the word has not been loaned into the Marshallese language. Marshallese words similar to the English term *environment*—*peḷaak* and *meḷan*—simply refer to one's surroundings; Marshall Islands never cite the *peḷaak* or the *meḷan* as the reason or cause of something, so they could not blame climate change on these entities. They could consider climate change to be natural without having any word for *nature* or *natural*—as in the case of the December 2008 flood (Rudiak-Gould, in press)—but they do not. For Marshall Islanders, noncyclical change cannot be natural. There is a right way for the seasons to be, just as there is a right way for society to be. In the case of the seasons, that right way includes a certain amount of change, but it is cyclical change, itself a form of stasis (Rudiak-Gould, in press). Thus Marshall Islanders would not consider long-term changes in the environment to be natural. Blame is necessary.

Given the unpopularity of God as a target of blame and the impossibility of nature as such a target, it is human beings who must be at fault. The question becomes which human group Marshall Islanders choose to accuse. The debate over climate change culpability and responsibility is raging worldwide (see Hulme, 2009: 155–156; Lahsen, 2004). Journalistic, anthropological, activist, and indigenist literature usually indicts the West, the north, or the industrial nations for creating climate change (see, for instance, Barker, 2008; Crate & Nuttall, 2009c: 11–12; Degawan, 2008: 54; García-Alix, 2008; Tauli-Corpuz, 2009), invoking an innocence narrative of indigenous people and the poor. NGOs such as Oxfam have adopted this blame strategy (Radio New Zealand, 2008), as have climate specialists in developing countries (Lahsen, 2004) and many indigenous leaders (Jacobs, 2005; Lindisfarne, 2010; Shearer, 2011; H. A. Smith, 2007) and radical groups (Baer, 2011). Meanwhile, those who place climate change guilt on all human beings—not just westerners, northerners, and industrialites—are equally passionate in their emphasis: for instance, an American missionary I spoke to in the Marshall Islands who decried Marshallese hypocrisy for declaring that they are at the front lines of climate change while also blithely polluting their own country.

These disagreements cannot be traced to differing views of the causation of climate change. All of the individuals and groups above (as well as Marshall Islanders) agree on the following propositions regarding the causes of global warming: (1) All people in the world contribute to the greenhouse gas emissions that cause global warming; (2) Affluent citizens and industrialized countries produce the majority of emissions while smaller countries and poorer people produce far less. Since both facts are widely acknowledged and agreed upon, in a sense there ought to be no disagreement about climate change culpability and responsibility, no debate about which emphasis is "correct" (Douglas, 1992: 9; Sarewitz, 2004: 390). Yet that debate is not just live but indeed fierce. Among the reasons for this is that an individual or society may choose to downplay the first or the second fact even while acknowledging that both are true. And this choice of emphasis matters greatly, because the two options yield starkly different "moral readings" (Bravo, 2009: 277) of climate change: either as foolish humanity's self-destruction or as greedy industrialites' victimization of the powerless. I will call emphasis on the first proposition "universal blame" and emphasis on the second proposition "industrial blame." As we will see, Marshallese citizens (though not their government) lean heavily toward universal blame, with a special emphasis on their own culpability (in-group blame). Understanding the causes and consequences of this perhaps surprising perception is the aim of this chapter.

MARSHALLESE NARRATIVES OF CLIMATE CHANGE RESPONSIBILITY

In September 2000, the *Marshall Islands Journal* printed a political cartoon (Figure 4.1). Many human artifacts are visible—missiles, cars, airplanes, factories—and Marshall Islanders associate these technologies more with large industrial nations than with their own "small island state." But the cartoon pays no attention to this disparity. Instead, the caption reads, "People [*armej*] are really burning the world with poison." It does not claim that "Americans" or "people in the big countries" are doing so but that "people" in general are doing so.

At the Women United Together Marshall Islands (WUTMI) forum on climate change in April 2009, after presentations that highlighted the threat that Styrofoam poses to the climate and the environment in general, an attendant raised her hand and commented, much to the amusement of the audience, that the presenters were themselves, at that very moment, using Styrofoam cups. The presenters were put on the spot. But rather than replying that their contribution to climate change was negligible, they showed signs of embarrassment. One of them tried to quietly slip his Styrofoam cup out of sight. Later, the women in attendance decided upon six recommendations for climate change response. None involved protest. Instead, they involved local adaptation and mitigation: the reduction of the Marshall Islands' own contribution to global warming through such measures as recycling, cutting the use of air conditioners, and encouraging public transport.

In response to a 2004 article about then-President Kessai Note's protests in international fora about the carbon sins of industrial nations, a

Figure 4.1 Cartoon in the *Marshall Islands Journal*, September 15, 2000. (Reprinted with permission of the *Marshall Islands Journal* [www.marshallislandsjournal.com].)

Marshallese man posted the following on Yokwe.net: "We keep hearing the same complaints from our government officials when the[y] speak in forums and assemblies around the world and when they come home what we see them drive are trucks, SUVs, and cars. . . . All use gas."

In 2007, another Marshall Islander on Yokwe.net chimed in:

> I don't know where the president is getting his facts from. The real threat to our environment is our own activities. Looking around Majuro, the real threat our environment faces everyday come from the locals and businesses themselves. Until our leaders do something about these real threat[s], what they are telling the rest of the world is not real. . . . Tell those nations the real story at home and stop making good faces. We are the real problem here.

On Yokwe.net in 2008, a Marshallese immigrant to the United States wrote another comment in response to a posted article on climate change:

> We have not been keeping our beaches and our land clean and sanitized. We also play a good deal in what is to become the destruction of our beloved islands. . . . One way that our islands could be saved is to stop polluting the sea. When we leave our trash *ilo lik* [on the ocean beach], it kills the barrier reef that's protecting our island. We have to do something about it and fast. We are also responsible for what is to become . . . the destruction of our beloved islands.

Other Marshallese Internet users wrote in to express their agreement.

The Marshallese government's Office of Environmental Planning and Policy Coordination (OEPPC) held a climate change poem competition in 2010. The winning entry, by Crystal Kabua, expressed no outrage at foreign groups. The focus was instead on the burgeoning local impacts of this "long, slow terror." In the last stanza, culpability was assigned: "We are the ones causing climate change."

During the staged student debate between the University of the South Pacific (USP) and the College of the Marshall Islands (CMI) on resettlement as a response to climate change, even the reality of climate change was open for dispute. But culpability was not. The CMI debater, arguing that climate change could be stopped and therefore resettlement averted, took in-group responsibility as a given: "Instead of using fuel, we can use solar. That's one way [to fight climate change]. We can stop dumping our garbage into the ocean." The statement went unchallenged.

When a reporter from the *Marshall Islands Journal* asked students in Majuro what should be done about global warming, none of them mentioned campaigning abroad or protesting against industrialized countries. Instead their answers were the following:

> What we can to do about this is . . . not burning chemicals, and not helping the tides take the pieces of our small islands. Like taking soil from beaches or dredge sites so much for building houses.
>
> We could help clean up the island.
>
> We the people are the ones causing global warming. If we would really care about it, then we would have at least done something to prevent it from happening. (Journal, 2008b)

High school-aged students at Assumption School in Majuro, as part of their participation in a climate change-themed summer program in 2012, made posters on climate change. One question they were expected to answer was, "How do we convince others to take action to help slow global climate change?" While the word *others* could easily have been interpreted as foreigners, Americans, or affluent citizens of large industrialized nations, it was instead interpreted without fail as other Marshall Islanders or other people in general. The students advocated planting trees, burning less fossil fuel, conserving resources, eating local food, and riding less often in cars. The students were later asked to make short videos encouraging climate-friendly behavior. The videos advocated local behavior changes: energy-efficient light bulbs, carpooling, recycling, turning off lights, and eating local food.

When I asked a Majuro man why the *mejatoto* was changing, he said, "Because of chemicals. These things that come from power plants." He then gestured to the *Majuro* power plant. Another man in Majuro said that crops had not been growing as well as before because of the heat, and that this was because of cars. "Where are these cars?" I asked, and he named the most developed *Marshallese* communities—Majuro, Ebeye, Jaluit, and Wotje—and then added, "We used to use lanterns. Now we use generators. This harms our lives." Even when it is admitted that industrial countries possess far more of these climate change causers, many locals argue that the only reason that this is the case is because Marshall Islanders are poor; if they were as rich as First World citizens, they would pollute just as much.

Rien Morris, the senator from Jaluit Atoll, answered without hesitation in 2009, when I asked him who is responsible for climate change, "We are, you and I. Marshallese may not think they are responsible for it, but they are. It's only a little contribution, but we too have air conditioners and cars. All Pacific islands contribute. . . . We are in it together." To gauge his reaction to a different blame strategy, I then told him that some people argued that *Americans* cause climate change. He was unconvinced:

> No, *everyone* does. If you go dive here, you will see all the plastics. Marshall Islands is only a little, but, by joining with all the other Pacific Islands, it makes a big contribution. All the batteries and plastic—they all contribute. . . . The pollutions from the Marshall Islands are also a contribution to all these problems which will affect the environment.

At the Assumption School climate change symposium in June 2009, I asked a few attendants if locals should feel angry about climate change. A woman said, "There *is* anger," but then directed that anger at the Marshallese government, not foreign countries: "For instance, *I* get angry—at the government here for allowing people to keep dredging and things like that."

Those who mourn the extinction of the Ripako and the erosion of the Bōn fault the clan itself for that loss. People abandoned their heritage land on Piñļap, forgot their clan membership, ignored their elders, and failed to produce heirs; they shirked their duty to safeguard good ancestral things for the next generation.

These examples are not cherry-picked. More systematic assessments come to the same conclusion. When asked point-blank whose fault (*an wōn bōd*) climate change was, only 18% of survey respondents named an outside group ("scientists," "those who went to the moon," "the big countries") while 52% blamed "people" (*armej*) in general or "us" (*kōj*) and 3% blamed Marshall Islanders specifically (5% "blamed" God, 26% said they did not know, and 6% said no one was to blame[2]). In a more oblique (and therefore probably more reliable) method of assessing culpability narratives, I asked survey respondents "What should people do about climate change?" Results were similar. Only 4 of the 26 suggestions for mitigating climate change were aimed at other, larger countries. Of the remaining 22 suggestions, 17 were general prescriptions for reducing pollution or protecting the environment (not specifying whether it was just Marshall Islanders or people in general who should do this) and 5 seemed to be aimed specifically at Marshall Islanders.

A smaller number of Marshall Islanders express industrial blame. A cartoon published in the *Marshall Islands Journal* in January 2010 shows a drowning man, labeled "Pacific Island Nation," saying, "You're the one who made it this way . . . help!" (*Kwe eo kwar kōmman bwe en eindrein . . . jiban!*) to a man on the shore labeled "Bigger Nation." Bigger Nation replies, "That's because I love money . . . but I'll build you a seawall" (*Kōn aō yokwe jāān . . . ak inaaj kōmman am seawall*). A middle-aged woman on Ailinglaplap blamed America for climate change when I spoke to her in 2009, despite my suggestion that some Marshall Islanders disagreed with her:

PRG: Will Marshall Islanders stay in this country in the future?

X: I don't know if they will, because there will be a flood. I heard this on the news. The ocean will rise ten or more feet. The problem comes from America and places like that, not from here. It's because the ice is melting.

PRG: Why is the ice melting?

X: Because the *mejatoto* has been damaged, because of Americans' tools. . . . They make war and ruin the *mejatoto*.

PRG: Why are Marshall Islanders not angry about this, if Americans are causing it?

X: I *am* angry. When the ice melts, where will I go? [Angrily] We have no power.

PRG: What about God's promise not to flood the earth?

X: That's true—it won't flood again like in the time of Noah, because of people's sins. But now the Americans are ruining things. Americans are doing it. They are destroying the good things that God made.

PRG: Some Marshall Islanders says that *all* people contribute to the problem, even people in this country. What do you think?

X: No. Marshall Islanders may damage their reefs with plastic imported from America, but they're not causing the sea to rise.

In an even more vociferous statement of industrial blame, a young Marshallese man named James Bing spoke at the Global Humanitarian Forum in 2008. He performed a furious Marshallese war chant and dance and then said, blaming industrialized nations not just in the abstract but in the flesh, "This is how angry I am. Rising sea levels have taken our sand, our beaches, our trees, our food, and most importantly, our soil. Where is my soil, ladies and gentlemen? What have you done to it? I want my soil back." For one government official, the same man we encountered in Chapter 2 as a "hyperconvert" who believed that a small islet in Majuro Atoll had disappeared owing to climate change, "an entire island destroyed" had become a privately voiced protest for both nuclear testing and climate change. Such individuals put climate change into the framework of nuclear testing as an evil force inflicted by American arrogance and negligence. But these individuals are greatly outnumbered by those who do not.

THE ORIGINS OF IN-GROUP BLAME

I will argue that the popularity of universal or in-group blame for climate change stems from the threat's framing in terms of modernity the trickster, with its moral focus on the tempted rather than the tempter. But lest this explanation seem too easily arrived at and lest Marshallese self-accusation seem to be accidental rather than systematic, rhetorical rather than genuine, pragmatic rather than idealistic, it is essential to consider various other explanations.

Could it be that Marshall Islanders have never considered the option of blaming outsiders, and simply default to in-group blame? Certainly not. Locals are perfectly familiar with the narrative of industrial blame. When Marshallese government officials air their climatic plight abroad (often as a member of the Alliance of Small Island States and the Climate Vulnerable Forum) they adopt industrial blame, contrary to most of their citizens. Charles Paul, the charge d'affaires of the Marshall Islands Embassy in the United States, testifying in 2008 before the US House of Representatives Foreign Affairs Subcommittee on Asia, the Pacific, and the Global Environment, stated:

> The RMI has virtually nil GHG emissions on a global scale; small island developing states such as RMI contribute the least to causing climate change, yet remain the most vulnerable to its impacts. . . . The RMI is concerned about the international responsibilities of all major emitters. . . . The production of each ton of CO_2 is a small assault upon our shores.

At the Copenhagen climate summit in 2009, then-President Jurelang Zedkaia told world leaders and other attendees:

> Distant promises are no longer enough; my people, and the world, deserve more. All I ask of you in this room is that you allow this generation of Marshallese, and generations to come, to survive as a people inhabiting a country that is but a collection of tiny specks of corals in the vast expanse of the Pacific Ocean. They are the lands of my ancestors. Allow these tiny specks to sink beneath the waves, and you will have destroyed an entire nation, a culture, a people. I will never be a silent witness to such an international crime. (Journal, 2010b: 5)

More recently, at the United Nation's 2012 Rio+20 Conference on Sustainable Development, President Christopher Loeak's National Statement described climate-vulnerable nations such as the Marshall Islands as "victims" and censured the "largest emitters" for "evading responsibility" (Loeak, 2012), and he began investigating the feasibility of forcing polluting nations to clean up based on legal precedents that make states liable for cross-border pollution.

Many Marshallese citizens are aware of these governmental narratives. Indeed, government representatives have transmitted the message directly to people. Recall the WUTMI climate change forums narrated in the previous chapter. While the WUTMI-affiliated organizers of these conferences tended to endorse universal blame, they also invited speakers from the government who advocated industrial blame. Attendees heard both blame narratives in the space of a single educational session. At the WUTMI April 2009 forum, Yumi Crisostomo emphasized industrial blame, but later an audience member reprimanded the presenters for their own use of Styrofoam, implying universal blame. Sometimes both blame strategies were advocated by a single individual: at the WUTMI climate change forum in July 2009, an EPA official, Michael Honeth (not himself Marshallese but a representative of the Marshallese government) told the attendees *both* that the RMI was too small to contribute to climate change *and* that Marshall Islanders ought to reduce their carbon footprint and reconsider certain maladaptive local development practices. So Marshall Islanders are familiar with industrial blame. When I suggested industrial blame to interviewees, they were never surprised; they simply rejected the suggestion.

Could it be that there is no convenient category for "industrialized nations" in the Marshallese language or in Marshallese minds? No. Marshall Islanders have a phrase that corresponds roughly to what we might call high-income, First World, large, or industrialized countries. This phrase is *laļ ko rōļļap*—literally "the big countries"—with the Marshall Islands often referred to, in contrast, *āne jiddik kein ad*, "these small islands of ours." For good or ill, Marshall Islanders have internalized the idea of smallness, whose most recent manifestation is the category "small island state" (see Hau'ofa, 1993). Largeness is here associated with power and wealth and smallness with powerlessness and poverty; in a presentation at the WUTMI executive board meeting in 2009, a presenter referred to "the small countries [*laļ jiddik ko*], those countries that have few resources, unlike countries like America and other big countries [*laļ ko rōļļap*]." Locals also have terms for "American" (*riamedka* or *ripālle*) and "America" (*Amedka* or *aelōñin pālle*), terms that are used frequently, laden with associations, and possessed of a long and complex history, as we saw in Chapters 1 and 2. Marshall Islanders are perfectly able to indict *laļ ko rōļļap* or *Amedka* and exonerate *āne jiddik kein ad*, but they choose not to.[3]

Do Marshall Islanders adopt in-group blame because they do not appreciate how small their contribution to climate change is compared with other countries? One could invoke Epeli Hau'ofa's argument that the notion of the Pacific as a group of "small islands" is a belittling western fantasy at odds with indigenous conceptions of Oceania as large (Hau'ofa, 1993). One could point to the fact that Marshallese outer island children often speak of the United States as a collection of small islands or atolls called California, Arkansas, and so on, and assume that Americans whom they have met must be acquainted with or related to each other. In many other ways these children show that they conceive of the United States and the Marshall Islands as being similarly scaled. Marshallese adults, while much more aware of America's large size, are nonetheless often astonished to hear that America has 300 million citizens. An anecdote illustrates this:

> One Marshall Islands *iroij* [chief], on his first visit to Guam, which measures two hundred and nine square miles and has a population, including American servicemen, of seventy-odd thousand, asked to see the sights. He was taken through Agaña, a modest town with few buildings more than two stories high. He was shown a Strategic Air Command base. . . . He also got to ride in an elevator. After all these experiences, he was moved to exclaim, "Tell me, is America as big as Guam?" (Kahn, 1966: 224)

Perhaps an underestimate of the size of foreign countries—a worldview in which Marshall Islanders form perhaps 1% of the world's population rather than the true percentage of less than 0.001%—may nudge locals away from industrial blame. But surely it can do no more than nudge them, because Marshallese adults *are* aware—acutely aware—that countries like America, China, and Japan are much larger than the Marshall Islands. Well-educated interviewees who could recite America's exact population nonetheless chose to adopt universal blame. A middle-aged educator in Majuro acknowledged that the bulk of emissions originated in foreign countries, but then said:

> Let's picture a blank white sheet of paper. And each country would come with their big brush of paint and paint it, a . . . black spot on

> the white paint. . . . The Marshall Islands with its small part as a contributor to . . . global warming, take this tiny part and put a little dot in that white sheet of paper. That is no longer a clear white sheet! It's tarnished, regardless of how small that thing is. So we do have something to contribute to the whole picture. . . . Even though it's small it's still something. . . . It is also our problem. And even though we're only contributing a little, we're affecting at least somebody.

In an editorial in the *Marshall Islands Journal* in 2007, the writer acknowledged that "The Marshall Islands are only a speck of dust in the world compared to the US, Asia, Europe and Africa, where most of the global warming [e]ffects are coming from," but then adopted universal blame nonetheless:

> We, on the other hand, are contributing to global warming by cruising around in our cars, throwing garbage all over the place, using tons of tin foil to take out food after kemems [first birthday ceremonies], and using chemicals that also contribute to the effects. . . . Every time you throw something on the ground, drive around and waste gas . . . use a whole box of tin foil for one plate, think about where you're going to call home in the next 30 to 50 years. (Bigler, 2007)

Another explanation based on ignorance might hold that Marshall Islanders adopt universal blame because they do not understand the physical causation that leads from foreign emissions to local climate change. Recall from Chapter 2 that only about a third of my survey respondents could provide some account of the scientific cause of climate change. Some locals explain the causation of climate change in ways that a scientist might regard as preposterous: the sun has moved closer to the earth, the ozone hole is letting in more warmth, the atmosphere is like a torn blanket. One local thought that the greenhouse effect was caused by greenhouses. Scholars have observed that, like so many hazards in "risk society" (Beck, 1992), the causation of global warming is tremendously difficult to grasp, originating from countless sources across the globe, occurring over huge time lags, involving numerous intervening steps, and requiring knowledge of such arcane subjects as atmospheric chemistry and arctic ice dynamics (Kasperson & Kasperson, 2005[1991]; Sterman, 2011; Ungar, 2007: 83). Even the one third or so of Marshall Islanders who could give some account of this chain of causation often required numerous leading questions in order to trace the causation back to its source—many interviews proceeded as follows:

PRG: Why is erosion occurring?
X: Because the sea is rising.
PRG: Why is the sea rising?
X: Because the ice is melting at the North Pole.
PRG: Why is the ice melting?

X: Because the world is getting warmer.
PRG: Why is the world getting warmer?
X: Because there is too much sun getting in now.
PRG: Why is too much sun getting in now?
X: Because the blanket up there is torn.
PRG: Why is the blanket torn?
X: Because of the smoke.
PRG: Where does the smoke come from?
X: From cars and factories.

One could argue that the reason that Marshall Islanders do not blame outsiders for climate change is that most of them do not understand how foreign countries could cause these local harms, or, even if they do, they find the chain of causation too long, the link between perpetrator and impact too tenuous. One could then cite a number of scholars who have argued that these difficulties of climate change causation, and therefore of climate change ethics, undermine people's capacity to cast blame for the problem (Alicke, 2000; Einhorn & Hogarth, 1986; Jamieson, 2007: 475–476; Sunstein, 2006b; Unger, 1996: 24–53), or that one of the main drivers of climate change apathy is the complexity and counterintuitiveness of the threat's causation (Bostrom & Lashof, 2007: 38–39). In-group blame, by this argument, is appealing because it is easier to conceive of.[4]

But this is a faulty line of reasoning. Understanding the minute details of causation may strengthen a person's blame, but it does not create it. In the theory of risk society, Beck (1992) and Giddens (1999, 2002) at times imply that the proliferation of diffuse, invisible, "causally amorphous" (Unger, 1996: 48–49) risks gives birth to a society in which no one can be held accountable for anything (Beck, 1992: 32–33; Giddens, 1999: 8). But it is also true that the same mysteriousness of causation renders blameworthiness *easier* to assign, since the range of possible perpetrators and plausible causal links is greatly expanded, as Beck himself also notes (1992: 27–28). Moreover, both ethnographic and experimental research indicates that when people wish to accuse, they will find a way to do so: harm arouses the desire to identify a perpetrator (Alicke, 2000: 569; Robbennolt, 2000), and people do not demand to know all of the intervening steps of causation as long as the identified culprit is ideologically acceptable:

> We do not have to wonder how people come to believe in the mysterious connections. . . . Plausibility depends on enough people wanting to believe in the theory, and this depends on enough people being committed to whatever moral principle it protects. . . . A community censors its own . . . causation theories. Those which no one wants to use can be easily discredited. Those which a large enough category of people find it convenient to use will get acceptability. (Douglas & Wildavsky, 1982: 38–39).

This observation is borne out in many studies. Byg and Salick (2009) provide a case study of Tibetans who confidently blame climate change on humans despite being hesitant, unsure, and conflicted in their explanations of how exactly this occurs. In particular, when an agent is considered all-powerful, people neither need nor expect to understand how that agent causes what it causes. In the Marshall Islands, such agents include God, radiation, and (to a certain extent) America and scientists. Locals confidently attribute a variety of phenomena to radiation despite being unable to provide an explanation for how such effects are produced. Radiation's mysteriousness, in fact, makes it easier to accuse, as people do not expect to understand how it creates its harms. The same is true of climate change, as I discussed in the previous chapter. Marshallese public understanding of the scientific causes of climate change is indeed hazy, but that does nothing to stop people from blaming whomever they choose.

Marshall Islanders might have a number of reasons to point the finger at America or other large countries. They consider these nations far more powerful than their own. They often credit foreign scientists and Americans with extraordinary powers. If climate change entails the destruction of entire islands, could Americans cause such a thing? Certainly—they already have, in the era of nuclear testing, and much more quickly and dramatically. A humorous anecdote told by Marshall Islanders demonstrates the powers with which locals credit Americans. Some Marshall Islanders were driving a motorboat at top speed, late at night, in the Kwajalein lagoon. A few of the passengers were worried that they might hit a reef or island in the dark. But there was an elder aboard, and he said, "Don't worry! I know this atoll in perfect detail." Just as he said that, the boat hit a coral reef, flew into the air, and came to rest on dry land. The elder said, "Damn these Americans—now they're even moving the coral!" Marshall Islanders might also have deeper motives for accusing America. Notwithstanding some positive discourses and attitudes about Americans, Marshall Islanders also criticize American culture as the degenerate mirror image of Marshallese tradition. While Marshallese outrage over nuclear testing is generally much more muted than outsiders expect, it does exist (see Barker, 2004: 99–102; Dibblin, 1988: 37, 52; Niedenthal, 2001: 16). Marshall Islanders could point out parallels between climate change and nuclear testing as acts of criminal negligence committed by America upon a blameless Marshall Islands, but few of them do. The question remains as to why.

Expatriate Americans often give a particular answer to that question: Marshall Islanders shy away from other-blame because theirs is a culture of nonconfrontation. They do not wish to cast any stones. There may be some truth to this. Locals often emphasize kindness (*jouj*) and respect (*kautiej*) as keystones of *m̧anit*, and some Marshallese proverbs encourage composure and acquiescence: *Illu, luuj* ("He who gets angry loses") and *Illu, m̧ōk* ("He who gets angry simply tires himself out"). Western visitors to the Marshall

Islands often perceive the Marshallese as hospitable, forgiving, and meek (Walsh, 2003: 386), and some scholars agree: Francis Hezel points to Micronesians' "infinite capacity to shrug off . . . wrongs" (2009b: 4; also see Barker, 2004: 99; Spoehr, 1949: 90), a habit purportedly stemming from thousands of years of cultural evolution on small, crowded islands where resentments must be swallowed rather than spat out. On Ujae, I observed a pattern of withdrawal or anonymous sabotage as a response to public conflicts (Rudiak-Gould, 2009b: 209–211).

But a reluctance to rock the boat in small, closely knit communities can hardly explain why Marshall Islanders would choose to turn climate change blame inward, toward the very people on whom their daily well-being most directly depends. Moreover, the peace-loving, forgiving ethos of Marshallese society should not be overstated. The notion verges on a colonialist discourse of native meekness, and it ignores contrary evidence. How are we to explain the chronic warfare (Spoehr, 1949: 90) and rampant chiefly assassinations (Kiste, 1974: 4) of precolonial times? Or, more recently, protests against the American expropriation of Kwajalein Atoll (Dibblin, 1988: 98–100), Compact of Free Association negotiations (Journal, 2003a), nuclear testing (Barker, 2004: 23–24; Dibblin, 1988; Kiste, 1974), access to a dock in Majuro (Journal, 2008a), and violence against women (Journal, 2010i)—not to mention open insults and violence directed at Chinese immigrants, or the shouting matches and fistfights that occasionally but memorably break out in outer island communities (see Carucci, 1990)? Nonconfrontation cannot get us far in explaining the lack of industrial blame for climate change.

Perhaps Marshall Islanders fault themselves for climate change because they have been the victims of "symbolic violence" (Bourdieu, 1994): they have been duped into a false consciousness of in-group accusation, blinded to the international injustice inherent in climate change. Perhaps the wider pattern of in-group blame in Marshallese society stems from a long history of "symbolic domination," but it is highly unlikely that in-group blame for climate change per se stems from it. Anthropological studies of symbolic domination document the process at work in the here and now via observable communications (see for instance Lyon-Callo, 2000). But in the Marshall Islands no one with an obvious vested interest in mystifying environmental injustice is advocating in-group blame. In-group blame's public advocates are various leaders of WUTMI, a local NGO dedicated to improving the lives of Marshallese women and the communities they support; they have no obvious reason to blame the victim. Even if they were for some reason tempted to do so, it would be an odd strategy considering that the WUTMI spokespeople are themselves Marshallese, bearing their own share of the guilt they promote.

Perhaps Marshallese climate change communicators are choosing in-group blame strategically. It has certain practical benefits as an educational strategy. If the communicators tell the public that "the big countries" are the real movers and shakers of climate change, the result can only be an overwhelming sense of disempowerment: a tiny country with no political clout could not possibly sway the decisions of the world's mightiest nations. If the communicators instead portray the Marshall Islands as a carbon contributor, the opposite will result: empowerment is the positive flip side of guilt (Minnegal & Dwyer, 2007: 47–48). And that sense of empowerment is absolutely essential for successful climate change outreach. Protection motivation theory (Maddux & Rogers, 1983; Rogers, 1975), a psychological theory with extensive empirical support (Floyd et al. 2000; Grothmann & Reusswig, 2006) and demonstrated on-the-ground relevance to climate change responses (Grothmann & Patt, 2005; Kuruppu & Liverman, 2011), predicts that when people do not feel efficacious and do not believe that available protective measures are effective, they will not act to protect themselves. More than that, they are likely to doubt the very reality of the threat (Lehman & Taylor, 1987; also see Barlett & Stewart, 2009: 356–359), justifying this comforting denial through whatever discourses are available (Grothmann & Patt, 2005: 203; Grothmann & Reusswig, 2006: 106; Kroemker & Mosler, 2002; Swim et al., 2009: 126). As two more ethnographically oriented scholars write:

> If available knowledge is useless, or even (socially) dangerous, there may be no point in taking on the often considerable costs involved in assimilating it. . . . [People] calculate how much understanding they are willing to own. . . . This often hidden process may then be recorded, misleadingly, as simple ignorance or resistance. (Jasanoff & Wynne, 1998: 41)

For those Marshall Islanders who feel that nothing can be done about climate change and who therefore wish to deny the reality of the threat, a convenient justification is at hand: the Genesis argument. Universal or in-group blame, by conferring responsibility onto Marshall Islanders, may be a smart educational strategy.

This might explain why organizations like WUTMI often promote this blame strategy. (When I asked a senior woman at WUTMI why so many Marshall Islanders blamed themselves for climate change while many outsiders think that Marshallese contributions to climate change are nil, she responded, "It's as simple as this: We think we're important and they don't think we're important. We think we matter and they don't think we matter.") But this does not explain why so many locals accept it: they are not forced to blame themselves simply because some activists have encouraged them to do so. Moreover, if locals were adopting in-group blame strategically in order to boost their sense of empowerment vis-à-vis climate change, one would expect much higher appraisals of self-efficacy on the issue. Locals make no ambitious statements of being able to stop climate change, only of having the obligation to reduce their contribution to it. I asked those survey respondents who had heard of it "Can Marshall Islanders solve 'climate change'?" Of the 76 responding, only 25% said yes while 58% said no. (17% gave mixed answers.) A middle-aged woman on Jabwor, Jaluit, told me about a Canadian photojournalist who had interviewed her for an article on climate change:

> She asked me, "If the country is flooded by climate change, what will Marshallese people do?" I answered "We'll make a boat just like Noah—we'll climb aboard, and wait it out." Then the journalist said, "But this time the water won't go down again." So I said, "Then there is nothing that we Marshallese can do."

Activist expediency turns out to be an inadequate explanation for the popularity of in-group blame. There is also something of a paradox here to be considered: locals tend to blame themselves for climate change even though they tend not to believe that they can stop it. A majority espouse in-group blame while only a minority espouse the high sense of empowerment that in-group blame supposedly entails. We could simply say that, as universal blamers, they criticize their own contribution yet realize that they cannot, by themselves, solve the problem. But the question remains: if their own actions neither make nor break climate change, why not shift accusation to the larger culprits? Something quite powerful must be encouraging people to indict the in-group.

I argue that this powerful influence is the discourse of modernity the trickster. The cultural theory of risk, outlined in the Introduction, predicts that when a calamity occurs, "someone already unpopular is going to be blamed for it" (Douglas, 1992: 5). In the Marshall Islands, that "someone already unpopular" is Marshall Islanders themselves; in the discourse of seductive modernity they have a virtuous culture but are reproachable for having abandoned it.[5] These narrative origins of in-group blame are obvious in the way Marshall Islanders discuss the issue of climate change. Consider the student statements of climate change responsibility with which this chapter began. Climate change (like cultural decline in general) was caused by using foreign artifacts of modernity—the students mentioned cars, electricity, air conditioning, and televisions—and abandoning native things, like trees and manpower. Climate change (like cultural decline in general) should be addressed by relying again on native things: large ships could be replaced by *kōrkōr* (canoes), cars by walking, foreign fuel with local energy. Recall the statement of the woman in the "Human-caused harms" section in Chapter 2, speaking about the WUTMI climate change forum in April 2009:

> We learned that Styrofoam cups are bad. They were brought here from outside the country—they didn't come from here. People throw them away and it's bad . . . it damages the coral. There was nothing like

> Styrofoam before in the Marshall Islands—it came from other countries. . . . We have too many things from outsiders—like money. We don't rely on our own things anymore. We depend on outside things.

An elderly woman of limited education in Majuro had noted a change in wind patterns since the past, but had never heard of "climate change" or *oktak in mejatoto* and didn't know its cause. When I told her that the change in weather was caused by people burning oil, the idea immediately struck her as intuitive and plausible: "Oooo," she said (the Marshallese way of indicating that one understands well). "Because there didn't use to be things like that in the past. Air conditioners and things like that." She pointed to the air conditioner in the room. Many other examples come from the awareness-raising events narrated in the previous chapter and the efforts at local mitigation to be discussed in the next section of this chapter.

An important point must be made about some of the climate-friendly technologies that are advocated on traditionalist grounds, such as solar cells, coconut biofuel, wind turbines, and bicycles. Vis-à-vis *m̧anit*, these are anomalous objects. In one sense they are definitively *un*traditional: the precontact Marshallese did not manufacture or possess photovoltaic cells; they are imported from overseas and purchased with money. But in another sense they are traditional: they rely on local resources (sun, coconuts, wind, human power) and, after they are bought, they no longer need a steady input of cash in order to work. Mary Douglas observed that when a thing falls between the cracks of two categories, societies attempt to neutralize the conceptual threat by defining that thing as an anomalous case of just one of the two categories (Douglas, 2002[1966]: 48). With this in mind, the grassroots popularity of climate change mitigation in the Marshall Islands probably owes much to the fact that solar cells, coconut biofuel, wind turbines, bicycles, and the like have been defined as an anomalous case of traditionalness rather than an anomalous case of untraditionalness. Renewable energy is now much like the fiber glass fishing spears that are popular in the outer islands. Although they are imported from abroad, they facilitate the reliance on local resources, a keystone of *m̧antin m̧ajeļ*. Although they require money to buy, they ask for no more cash once bought. They ultimately lead away from, not toward, "living by money." As a result, no one ever complains about these foreign-made artifacts as violations of tradition. Exactly the same can be said of solar panels. Solar-powered flashlights, though also foreign-issued, are praised for helping in the subsistence activity of night fishing and for obviating the need to buy batteries.

The result is that images of renewable energy no longer seem out of place in portraits of tradition. At the University of the South Pacific climate change-themed summer science camp in 2012 (see next section), one group of students illustrated their poster in a binary fashion. The top half showed a traditional Marshallese thatched house, native trees, an unpolluted landscape, sunshine, along with several of the renewable energy technologies about which the students had been learning: a rocket stove, an aqua lens, and a solar panel. The bottom half showed a factory spewing smoke, an oil tank, modern buildings, a landscape devoid of trees and littered with trash, a dearth of sunshine. Another group's poster juxtaposed a modern building and smoke-belching car with a thatched Marshallese house attached to a solar panel.

I talked to all of the attendees of that summer science camp in a group, broaching exactly these issues of categorization. When asked, they all agreed that things like coconuts, wind, and waves were local while oil was not. (Sunshine took a bit longer to decide, but then the chorus answer—the answer given in unison by most of the group—was "yes, it is local.") They unhesitatingly answered no when I asked if any part of *m̧antin m̧ajeļ* caused climate change, and when I asked if renewable energy was local, there was some hesitation followed by a chorus answer of yes. They initially answered no when I asked if solar power was traditionally used by Marshall Islanders, but then, while discussing the localness of sunlight, they decided that solar power could be called part of *m̧antin m̧ajeļ*. When I asked if Marshallese people traditionally used wind turbines, they answered no, but when I asked if they traditionally used wind, they said yes, and a student gave the example of sailing. For wave power, a student gave the example of navigating using ocean currents. Another student pointed out that even ocean thermal energy conversion (a form of renewable energy derived from the difference in temperature between deep ocean water and surface water) was traditional, in a sense, because it uses warm seawater, just as Marshallese have always done when they bathe in the lagoon. In this group discussion, I observed students feeling out for themselves whether wind turbines, photovoltaic cells, ocean thermal energy conversion, coconut biofuel, and so forth could be called traditional, and they seemed to come out of it with the idea that these technologies were more anomalous cases of traditionalness than they were anomalous cases of untraditionalness.

If the traditionalist framing of clean energy is taken into account, the paradox I pointed to earlier—moral responsibility without a sense of empowerment—is resolved. People are not merely motivated to avoid risk (as protection motivation theory assumes), but to feel and appear righteous according to the standards of their society (Crawford, 1987; Goffman, 1959; Leary & Kowalski, 1990; Leary, 1995). This "moral motivation" may trump protection motivation, inspiring people to act even if they believe the action is futile, because the effort itself proves that one is a virtuous individual (Chess & Johnson, 2007: 227, 230; Crawford, 1987; Weinstein, 1987: 329). According to protection motivation theory, one will not act if one believes the action is futile. But protection motivation theory applies to *threats*; by putting a moral gloss on climate change it is no longer merely a threat, but also an *opportunity*: a chance to enact and display virtue. As such the efficacy of the actions is rendered irrelevant, the effort becomes an end in itself, and people may act on climate change even if they doubt they can prevent the threat. This is precisely the pattern one sees in the Marshall Islands.

THE MITIGATION MOVEMENT

These narratives of responsibility are not empty talk. Grassroots Marshallese society has given birth to a growing mitigation movement, following the premises of universal/in-group blame. The EPA-sponsored Environment Day in 2007 focused on reducing local carbon emissions by walking to work, turning off lights, and avoiding the use of Styrofoam cups and plastic bags (Journal, 2007a).[6] Energy Fairs in 2009 and 2010, organized by the NGO Marshall Islands Conservation Society (MICS) and various government offices, educated students about climate change and encouraged Majuro Middle School students to adopt energy conservation and renewable energy for purposes of both energy security and climate change mitigation; other similar workshops were planned for the city of Ebeye as well as the outer islands of Jaluit and Wotje (DeBrum, 2010). In March 2010, the College of the Marshall Islands organized a public discussion on climate change in the wake of the 2009 UN Climate Change Conference in Copenhagen. Speakers included representatives from both the grassroots (Angeline Heine from WUTMI) and the government (Yumi Crisostomo from the OEPPC). Heine, advocating local mitigation, emphasized that the Republic of the Marshall Islands was among the most intensive users of fossil fuels in the Pacific and asked the attendees "How can we ask the bigger nations for help, when we are [also] a contributor to climate change?" (Henry, 2010). In 2010 the Kwajalein Atoll local government sought and received a US$50,000 grant from the Global Environment Facility/New Zealand Agency for International Development Small Grants Program to boost its use of renewable energy, with the explicit goal of raising awareness of climate change and fighting it (Journal, 2010e). In May 2010, as part of an event encouraging healthy behaviors—in particular the reinvigoration of pandanus horticulture as a salubrious and customary practice—students and teachers at the Majuro Co-op School recast the growing of pandanus as a climate change mitigation measure: they rode through Majuro with a sign saying "Bōb [pandanus] Bandits save Mother Earth from the Global Warming Demons," pointing to the plant's ability to turn harmful carbon dioxide into harmless oxygen and to protect the shoreline from climate change-induced erosion (Journal, 2010f). In June 2010, nearly 200 Marshallese Boy Scouts spent a weekend on Enemanit Island in Majuro Atoll to learn about energy conservation, with an eye toward both reinvigorating tradition and mitigating climate change. The boys learned how to make *enrā*, traditional plates woven from coconut fronds, which Marshallese traditionalist discourse praises as emblems of both subsistence and conviviality: recall *Enrā bwe jen lale rere* ("Food basket so that we take care of each other") from Chapter 1. The participants advocated the replacement of imported plates with these local crafts on both traditionalist and environmentalist grounds; one group of boys stated, "If 80 percent of the world population used *enra* then the world would be free from carbon product[s] contributing to climate change" (Journal, 2010g). In July 2010, the Canadian expatriate Tamara Greenstone organized a camp in Majuro for students to learn about environmentally friendly renewable energy, and a science fair in which they presented the results to others. Although the event was organized by a foreigner, locals received it well, considering it both an educational opportunity and an action to staunch climate change (Heine, 2010). In December 2010, mayors from the country's Ratak chain of atolls requested US$51 million in climate change-earmarked funding from the United Nations for outer island solar energy and related climate change education (Journal, 2010k).

Two events in 2010 and 2012, hosted by the Majuro campus of the University of the South Pacific and sponsored by the OEPPC, sought to give high school students hands-on experience with renewable energy technologies and make them leaders in the next energy generation of the Marshall Islands. At these climate change-themed summer science camps, universal blame and local mitigation were everywhere implicitly advocated over industrial blame and foreign protest. One of the event organizers, Olai Uludong, explained to me that the science camp was intended

> to teach [the students] that climate change doesn't just mean sea level rise. It means [things like] turning off the light when you leave a room. . . . The key to resilience, whether it's adaptation or mitigation, is behavior change. . . . The goal is to expose kids to the science, so they know they have options, and know that mitigation is doable even for nonengineers.

In one activity, students were asked to compare the smell of a locally made soap with the smell of an imported soap from Hawaii; they were expected to notice, by smell, that the imported soap had required fossil fuels for its manufacture. In a previous summer science camp, Uludong had told students that the hot dog dinner they were eating was a contributor to climate change. At first confused, the students came to understand the connection once Uludong explained that the factory that made the hot dogs, as well as the ship that brought those hot dogs to the Marshall Islands, both used fossil fuels. Reportedly a student was so moved by this that he ate only uncooked fish for a week (Journal, 2011d).

During the 2012 summer science camp, the students spent the week visiting Majuro's solar installations, listening to visiting speakers talk about renewable energy, and building their own models of renewable energy technologies: energy-efficient stoves, small models of hydroelectric generators and windmills, a working solar panel, a solar water distiller, and a variety of easily constructed devices (an aqua lens, parabolic mirror, and solar oven) to harness the sunlight for the purposes of cooking. Posters on the walls (some made by previous students, others from an educational flipchart made by the South Pacific Applied Geoscience Commission) explained climate change, sang the praises of renewable energy and explained the climatic evils of fossil fuels. When I asked students what was good about renewable energy or about their particular projects, they said that it saves money, does not use electricity, reduces our contribution to the greenhouse effect, is environmentally benign, and does not violate tradition. The students made posters presenting the "do's and don'ts" of energy use: the "do" side portrayed biking, running, turning off taps, planting trees, solar panels, and an outer island, while the "don't" side showed air conditioners, running taps, a generator, trash, and a car. Another student-made poster portrayed "Today" as an island with cars and a factory, a cloudy sky, and huge threatening waves; below it was "The future," with bikes instead of cars, solar panels instead of factories, a traditional outrigger canoe in the lagoon, sunny skies, and calm seas; an alternate future was shown where cars were still in use and everything had been swamped by the ocean.

On the final day of the summer camp, students presented these projects to community members in an informal science fair. The event was well attended and the adults, even those of limited education, responded immediately to the message. Looking at a "do" vs. "don't" poster, a middle-aged female attendee pointed to the "do" side (showing a thatched house, trees, solar panel, energy-efficient stove and aqua lens) and said, "Oooo, the outer islands. You don't spend money and create disease." She then pointed to the "don't" side, with its smoke-spewing factories, and said "Majuro" with a laugh. The students proudly displayed their renewable energy models to the adults, pointing out their benefits for saving money and caring for the *mejatoto*. One group of students had been assigned to build a model of an efficient greenhouse and compare it to a model of an inefficient greenhouse, but they had, of their own accord, and quite ingeniously, reinterpreted the project to be about the greenhouse *effect* rather than proper greenhouse design. Their presentation of this project to the adults was a textbook case of universal blame and climate change as seductive modernity. What was supposed to be the efficient greenhouse, in which heat was effectively trapped, was now the GHG-suffocated present, and what was supposed to be the inefficient greenhouse, from which heat could escape, was now the pristine GHG-free past. One of the students said:

> This project shows "climate change": a change in the *mejatoto* and in life. It shows the difference between the present life (*mour in raan kein*) and the past life (*mour in etto*). This box, on the right, is the past (*raan ko etto*), when our lives were good. The sunlight could escape. This box on the left is the present (*raan kein*). It's bad, blocked with pollution. The sunlight can't escape. So there is less rain, more erosion and waves,

> more heat. It's hotter because of our use of manmade things. The *meja-toto* wasn't ruined in the past, so those problems didn't exist.

A visiting man asked, "Who makes that pollution?" The student answered, "Us people! It's from factories, like MEC [the Marshalls Energy Company, which runs Majuro's diesel power plant]. It's from burning plastic. We use a lot of plastic." Man: "How do we stop this?" Student: "Reduce the use of diesel." Man: "But then we won't have electricity." Student: "We can use solar—like that project over there [gestures to the solar cell student project]. It's *free*. You take it from the sun."

The discourse of modernity the trickster is everywhere evident here. Fitting the pattern of blame-without-efficacy, which can only be accounted for with the moral motivation I have discussed, at none of these events was it suggested that Marshall Islanders could stop climate change, but it was always argued that the *effort* was worthwhile. To be sure, maintaining Marshallese custom and fighting climate change were not the only reasons given for adopting renewable energy. The more commonly given reason was that renewable energy saves money. This may bring up a doubt: perhaps this grassroots interest in renewable energy is not so idealistically motivated, but merely a pragmatic effort to save money. But when it comes to Marshallese narratives of seductive modernity, this opposition between "pragmatic" and "idealistic" falls apart. Saving money is simultaneously practical and symbolic, since less "living by money" means more living by tradition.

As the local mitigation movement gains momentum, the country has seen minimal grassroots protest of other countries' contributions to climate change. James Bing's aforementioned protest at the Global Humanitarian Forum in 2008 involved only a single individual. In October 2009, Youth to Youth in Health (a Majuro-based NGO) and the Ministry of Internal Affairs organized a Majuro rally against worldwide greenhouse gas emissions as part of the Global Day of Action on Climate Change. But even here they emphasized universal blame (Journal, 2009b). A prominent member of MICS, Milner Okney, has stated publicly that the country's contribution to climate change was tiny and other countries were the primary culprits, but he does not emphasize this blame in his work with the NGO and has never sought to involve it in protest. The climate change-focused youth group Jo-Jikum helped to realize 350.org's Connect the Dots campaign in Majuro, and that event contained a note of protest. Youths donned diving gear and had their photograph taken under water while they held a banner proclaiming, "Connect Our Dots. Your Carbon Emissions Kill Our Corals." The photograph was broadcast on 350.org's website and elsewhere. Yet the emergence of the fledgling NGO Jo-Jikum does not herald the beginning of a protest-based grassroots response to climate change. In my discussions with key members of Jo-Jikum, including Kathy Jetnil-Kijiner, Desmond Doulatram, and Mark Stege, protest was never said to be core, or even tangential, to the organization's mission. The organizers described Jo-Jikum as being intended to raise awareness and encourage Marshallese youth to "take responsibility" for climate change, rather than to protest. When I asked, Kathy Jetnil-Kijiner maintained that even the Connect the Dots event had not been envisioned as a protest. While she would not rule out the possibility of Jo-Jikum's organizing a protest in the future, she felt that it would be quite a departure from the organization's usual modus operandi. The Marshallese government has lodged official complaints against foreign emissions, but these campaigns have not involved grassroots participation.

As suggested by the national government's cosponsorship of some of these grassroots mitigation efforts, the Marshallese political elite supports the transition to renewable energy. While they often adopt industrial blame in their climate change campaigns, they are also keenly interested in weaning the country off fossil fuels. Their reasons, however, may be partly distinct from those of the people. In June 2008, President Litokwa Tomeing's cabinet declared a state of emergency: with the price of oil spiking to nearly US$150 per barrel, importing fuel for Majuro's power plant had become cripplingly expensive. Since then the government has worked industriously to eliminate fossil fuel dependence. Funded by donors including the Asian Development Bank (ADB), the US Department of State, and Taiwan's International Cooperation and Development Fund, it has sought to completely electrify outer island communities using solar panels, use coconut oil as biofuel, install solar-powered streetlights in Majuro and a solar panel installation at Majuro's hospital that feed into the electrical grid, construct a waste-to-energy plant in Majuro, investigate the feasibility of an ocean thermal energy conversion plant in Kwajalein, and join international coalitions such as the International Renewable Energy Agency. Ambitiously, the government, under Jurelang Zedkaia, vowed to reduce the country's greenhouse gas emissions by 40% below 2009 levels by 2020, a pledge made in association with the country's endorsement of the Copenhagen Accord. These policies are justified on grounds of both energy security and climate change mitigation (see, for instance, Journal, 2010h). The mitigation movement that I have described could therefore be interpreted in more cynical terms as stemming not from genuine sentiments of climate change responsibility but rather from pragmatic economic concerns, which are strategically justified with environmentalist rhetoric. Indeed, government officers emphasize economic reasons as often as climate reasons. A particularly uncomplimentary interpretation might be that the government more generally feigns interest in climate change in order to woo foreign donors. Displaying climate change concern has opened up funding channels from the Asian Development Bank (ADB), the US Department of State, Taiwan's International Cooperation and Development Fund, Japan, and the Grand Duke of Luxembourg, and may in the future win donations from the World Bank, the Green Climate Fund, the Clinton Foundation, the European Union, Finland, Japan, and the United

Arab Emirates. A debt-for-adaptation swap is now being pursued, promising to relieve the government of over US$50 million of debt to the ADB (Journal, 2012b). One might also argue that the mitigation movement is not a product of guilty feelings, but rather a calculated attempt to shame larger countries into action or to demonstrate the feasibility of carbon neutrality; spokespeople have indeed sometimes described the government's motives in these terms (see, for instance, Journal, 2010b; Loeak, 2012). There may be some truth to these alternative explanations, but only in the governmental arena. In the grassroots realm, on which this book focuses, these alternative hypotheses are implausible. There would have to be a vast conspiracy, involving thousands of citizens, to feign in-group blame and interest in local mitigation—a far-fetched proposition. The grassroots mitigation movement is genuine, and its foundations are as ideological as they are pragmatic.

This movement is not yet fully grown. It will only be so once Marshall Islanders across the nation are regularly scolding each other in their everyday lives for driving cars, traveling in motorboats, and the like. Whether it will grow to this point is unclear, but there are reasons to think that it might. Recall the incident at the WUTMI climate change forum in April 2009, in which a woman chastised the speakers for hypocritically using Styrofoam cups. This shows that the mitigation movement has begun to operate on the micro-scale, with enough moral force to cause embarrassment and perhaps to inspire behavioral change. Built on a solid ideological foundation, the movement has potential to mushroom into a societal force truly to be reckoned with.

CONCLUSION

Marshallese approaches to climate change responsibility may puzzle and distress environmental justice advocates, presenting an awkward challenge to the notion of an indigenous counterhegemonic narrative of climate change (see H. A. Smith, 2007) and an uncomfortable counterpoint to the activist trope of frontline victims and innocent islanders. Moreover, if my analysis is correct, this problematic or anomalous viewpoint cannot be domesticated through dismissive explanations based on ignorance, rhetoric, deceit, or Canute-like delusion, for the attitude stems from deeply held moral convictions. At the same time, if my analysis is correct, the Marshallese response is not as disturbing as it may first appear if only we are willing to see it for what it includes and not just for what it lacks. It transforms climate change into an opportunity: a celebration of Marshallese distinctiveness against the forces of homogenization, the ultimate argument against blind modernization, an attempt to *use* climate change rather than *solve* it, as Mike Hulme advocates (Hulme, 2009: 328). If it is not exactly "critical," it *is* counterhegemonic in another regard: as a nongovernmental, morally motivated, unilateral movement against climate change, it powerfully undercuts the dominant paradigm in climate policymaking, namely that the problem must be addressed at the level of states (H. A. Smith, 2007), that it will only be addressed out of self-interest, and that therefore no meaningful action can be taken by any one actor until all others have also committed to acting. In a deep sense it is postcolonial as well: self-censure animates Marshall Islanders not out of a colonial inferiority complex but because they see in it the possibility of saving that which imperial intervention has threatened.

5 Modernity's Second Coming

The Unsettling Issue of Resettlement

When President Anote Tong of Kiribati announced that his people might someday have to abandon their low-lying island nation, *The Independent* transmuted that cautious statement into a more definitive headline: "Paradise lost: Climate change forces South Sea islanders to seek sanctuary abroad" (Marks, 2008). The possibility of future exodus had been reported as the reality of present exodus. The slippage of tense and mood was probably not accidental, because western audiences, when they think at all about the climatic plight of low-lying archipelagoes, imagine them only as "disappearing island nations" or "canaries in the coal mine" that must die (or have already died) in order to warn us (Farbotko, 2010). Relocation is assumed to be inevitable—the only response that such a marginal, powerless population could take (Barnett & Campbell, 2010; Mortreux & Barnett, 2009). But mass resettlement is climate change's most problematic "solution." It creates rather than averts the worst-case scenario. For Marshall Islanders, it is not clear that such a move would be preferable to drowning.

On June 19, 2009, several hundred Marshallese youth filled the pews of the Uliga Protestant Church in Majuro. This was not an ordinary church service: the youth had arrived from every corner of greater Marshalldom—each inhabited outer atoll, both urban centers, and immigrant communities in Guam, Hawaii, California, Oregon, Arizona, and Oklahoma—to participate in the two-week Youth for Christ National Youth Convention, with the theme "The Wind of Change." Since June 14 they had gathered each day at the church to take part in a variety of events, including lectures on youth issues. This day's lecture was delivered through an interpreter by a Tongan man named Makoni Pulu who worked for the Pacific Council of Churches in Fiji. This was a rare opportunity to speak directly to a wide swathe of the nation's youth, and Pulu sought to make the most of it.

"The wind of change has come," Pulu declared. He named urbanization, youth suicide, Chinese immigration, and western influence. But that was merely the preamble. He reminded his audience of the lecture he had given the day before, in which he described the threat of global warming and sea level rise. He offered three possible responses. The first was constructing seawalls. The second was planting trees along the shoreline. But, Pulu said, "These are not long-term solutions." The third response was relocation.

He began a slide show. "One of the biggest problems is denial," he said. "People don't want to move. People refuse to publicly consider the issue of resettlement." He showed a photograph of coastal erosion in Funafuti, Tuvalu, saying that it would be uninhabitable in twenty years and that it was only denial that allowed people to continue building settlements in such places. "The whole world is experiencing climate change," he said, and presented images of a flooded New Orleans. "China is experiencing climate change, too," he said, pointing to photos of floods. The continued habitation of low-lying Shanghai was an act of denial, as was the very existence of Nadi, Fiji. "Climate change affects agriculture, food, everything. But people don't want to move. The issue here is denial of the scientific evidence." He showed photographs of water lapping up on Fijian villages, of that country's 2007 floods, of king tides washing over islands in Kiribati, and, closer to home, of sandy beaches disappearing in Majuro. "There are places in the Pacific where people need to relocate," he emphasized. "Kiribati, Tuvalu, Tokelau, and of course the Marshall Islands."

He looked steadily at his audience, the youth of the nation, and said, "Are you ready to move?" The question hung heavily in the air. He continued, "Or relocate if this climate change comes?"

Marshall Islanders are experts at projecting superficial composure in trying situations, but this was perhaps too much. The attendees were visibly uncomfortable, fidgeting and nervously shifting in their seats. Pulu softened his tone. "Maybe I should change the question. Are you willing to move if this comes up?"

A member of the audience asked "Move where?"

"We'll talk about that next," Pulu replied.

The hard-line approach returned. "People are told to move, but they don't move—that's denial." He offered advice for "coming to terms" with resettlement, for "overcoming denial": one must show people the erosion and flooding that is already occurring and provide familiar examples of previous resettlements. He challenged the assembled youth: "Can you tell your elders at home to move to another country? It's important to provide information, to give options. You have to tell the story. People have already moved. . . . Do it face to face. Talk to Marshallese people who have already left, for example in Maui, about how their lives are, how resettlement is."

He then delved into the nitty-gritty of relocation: destination, timing, legality, costs, and cultural issues. Governments must help, he said, because they have money, and community leaders because they have influence, but in the end it comes down to individuals—everyone must be convinced. "In ten to fifteen years, Tuvalu will be gone," he forecast. "So it's not healthy to just keep standing while the islands sink. And New Zealand and Australia—they haven't exactly given the green light, but they're given the yellow light to the possibility of people from Kiribati and Tuvalu moving there. Of course, you'd lose all your coconuts, your crabs, your turtles. But those are the problems we're going to face. Do we have a choice? Can anyone here give me a choice of where to move?" No one answered.

"You have told me that you have experienced the sea rising up and moving into our islands. That is real. That is reality." He concluded, "I think there's only one thing to do about climate change, and that is resettlement. Thank you very much."

Visibly unsettled, the youth were now faced with the cruelly incongruous task of performing the light-hearted songs they had rehearsed during the previous days. But this new spectacle seemed to rescue the attendees from their despondency rather than grating against it. Four groups sang in turn, and the youth eagerly indulged in the clownery that so often accompanies such performances in the Marshall Islands. As a young man conducted his group of singers, he swayed his hips comically, to everyone's amusement. A man inserted himself between the singers and their jester-conductor to take a photograph of the latter, and the audience rewarded that jokey impudence with laughter. When the song finished the youth cheered and applauded furiously.

The events had concluded for the morning, and the congregation left the church. The singing had helped the youth regain their outward equanimity, but Pulu's speech still cast a pall. Taking an informal "exit poll" of the audience's reactions, I provoked nervous laughter and uneasy body language, and many of the youth I talked to found ways to dismiss Pulu's message entirely. A young man from Ailuk Atoll admitted with a strained laugh that perhaps some Marshallese might have to leave the country because of climate change, but then he reversed his position and declared that they could stay there forever. A man from Aur Atoll said that the speaker had been right, but then hastily changed the subject and hid his obvious distress with a smile. A married couple from Ebeye rejected the message outright. "I don't believe what he said," said the man.

> They say the water is entering the land, but it's not true. If we move to those other places with mountains, well, those same mountains make typhoons and hurricanes—whereas here in the Marshall Islands there are very few typhoons and hurricanes. My mother lives in Hawaii, but I'm staying in Ebeye. See how beautiful the Marshall Islands are?

He pointed to the image of an idyllic uninhabited islet on the outside of a water bottle made by Majuro's Pacific Pure Water Company. His wife was also unconvinced. "Scientists have seen climate change happening," she said, "but we don't really know. We'll just keep praying, keep praying. God said that the second time He destroyed the earth, it would be with fire, not a flood. We'll just keep praying."

Ultimately, Marshall Islanders voted with their feet. No movement toward evacuation resulted from Pulu's speech—not a single individual resettled as a result to my knowledge, nor was any eventual move planned for. Hoping to plant the seeds of mass exodus, he had succeeded only in ruining the afternoon of a representative cross section of Marshallese youth.

REJECTING RESETTLEMENT

It has been suggested in the literature that some atoll dwellers are unperturbed by the prospect of climate change-driven evacuation. John Connell submits that many Tuvaluan politicians have embraced climate change as an opportunity to win migration rights that they already desired for economic reasons (Connell, 2003). Carol Farbotko, while stopping short of suggesting that Pacific Islanders *wish* to leave, nonetheless questions the assumption that they conceive of themselves as tied to a particular territory, unable to effectively transplant their cultures, and culturally doomed in the case of climatic exodus: migration, she suggests, "is often a long-established and accepted characteristic of island life" (Farbotko, 2005: 288; also see Farbotko & Lazrus, 2012). Anthropologists often portray tradition, culture, and identity as endlessly pliable and reinventable (Camino, 1994; Jackson, 1994; Toren, 1999) and therefore far more resilient to foreign influence and territorial movement than we fear (Sahlins, 2005). Epeli Hau'ofa suggests that Oceanians yearn by their very nature to move and circulate; national boundaries are colonial impositions, and the true homeland of Pacific Islanders is not a particular nation state but the "sea of islands" as a whole, now including even parts of the Pacific Rim (Hau'ofa, 1993). Indeed, some Pacific Islanders have managed involuntary migration without major trauma (Knudsen, 1977; Larson, 1977; Tonkinson, 1977).

So is the hand wringing over "climate refugees" much ado about nothing? Perhaps islanders were expecting to leave their islands anyway, for other reasons. Perhaps they are staunch modernists who see in territorial abandonment a decisive end to traditional backwardness. Perhaps they see a financial silver lining to the cloud of deterritorialization. Perhaps they prefer to stay but regard their society as transplantable and therefore relocation as acceptable. Perhaps they have simply resigned themselves to the necessity, if not the desirability, of moving (indeed, the governments of every other atoll country have at least considered contingency plans for resettlement [Chapman, 2012; Keane, 2009; Russell, 2009; Schmidle, 2009; Toomey, 2009]). Such attitudes are not unimaginable in the Marshall Islands. While locals harbor negative discourses of America the trickster, they also give voice to positive images of America the savior and America the chief—why not take the opportunity to join the chief in his abode? Twenty-two thousand Marshallese citizens have, after all, already left to the United States for economic and educational reasons: there are so many Marshall Islanders in Springdale, Arkansas, that "Marshallese" is a box that can be checked on ethnicity questionnaires, and enough

Kili Island inhabitants in California and Arno Atoll inhabitants in Honolulu to inspire the nicknames "Kilifornia" and "Arnolulu." In the event of nationwide evacuation, it is likely that international donations would rain down on the Marshallese, these people for whom basic needs are the number one concern (Chapter 3). Might Marshall Islanders therefore see climate change exodus as an opportunity, good luck wrapped in bad?

They do not. Locals have so far, with few exceptions, roundly rejected the idea of relocation as a climate change response, despite various subtle and not-so-subtle exhortations to move or plan for movement. Pulu's speech was certainly the most strident suggestion, but it was far from the only one. In 2002, an American who had worked on Namdrik Atoll in the 1960s as a Peace Corps volunteer posted a message on the Marshallese online forum Yokwe.net in which he urged the Marshallese government to buy a 40,000 acre plot on the Big Island of Hawaii as a future homeland. At the Pacific Climate Change Roundtable meeting in Majuro in October 2009, Prof. Patrick Nunn from the University of the South Pacific in Fiji advocated evacuation on the grounds that it is simply necessary, a suggestion reported to the people through a front-page article in the newspaper entitled "Option for Sea Level Rise: Relocate":

> By 2100, I don't see how many islands will be habitable. We're now looking at a more than one meter sea level rise by the end of the century. . . . The biggest challenge is getting policy makers to understand the need for a profound change in the way Pacific people live. Relocation is one of the most difficult things to talk about and to convince people that the home they've lived in for centuries is no longer a viable option. There are no real options in Tuvalu, the Marshall Islands and other atolls other than to move people. Anyone looking objectively at this region has to see the need for relocation. (Patrick Nunn, quoted in Journal, 2009c)

A civil engineer visiting the Marshall Islands told students at the University of the South Pacific's 2012 climate change-themed summer camp, "What is certain is that the tide is rising here. This country will disappear—and there's nothing you can do about it." He went on to explain that seawalls were too expensive to build and insufficient even if built ("The sea always wins. Don't try to fight the sea. You will always lose."), and local mitigation was pointless because countries like China have power plants that do in one day the damage that the entire Republic of the Marshall Islands does in a decade. With both adaptation and mitigation dismissed, migration was the only remaining option.

Marshall Islanders have not heeded these foreigners' suggestions. Existing Marshallese migrants to the United States are not in any meaningful sense "climate change migrants," despite occasional suggestions in the media that they are (see Robinson, 2011).[1] People's stated motivations for migration are the usual suspects, shared by many other developing nations: the push factor of population growth at home and the pull factors of economic and educational opportunities abroad (see Pacific Institute of Public Policy, 2010)[2] as well as the existence of well-established Marshallese communities in Hawaii, California, Arkansas, and elsewhere. Nor has climate change evacuation been planned for, either formally or informally, at the individual, household, or community level. This lack of planning is not an oversight; it is deliberate. Most Marshall Islanders reject nationwide resettlement as a "solution" to climate change and refuse to resign themselves to it. When I asked survey respondents "What should people do about climate change?" (with 80 people responding and 82 individual suggestions made between those 80 people), only 2 suggestions involved relocation or planning for relocation. Compare this to 8 suggestions for adaptation; 26 suggestions for local or global mitigation; and 17 suggestions for continued study and discussion of the problem. Such responses do not reflect ignorance of the severity of climate change, since climate change is often spoken of as a dramatic rise in sea level, leading to the inundation of the entire archipelago.

All of the Marshallese NGOs involved in climate change have openly rejected resettlement as a solution. At the Women United Together Marshall Islands (WUTMI) climate change forum in April 2009, participating women expressed their disapproval of the idea of climate change evacuation, saying that they would lose *m̧anit* and be treated as second-class citizens if they lived abroad. The agreed-upon statement was as follows:

> Even though Climate Change may be out of our control the women feel relocation to a second home is not an option at this time. Instead they urge the government to bolster education system to help prepare and build resiliency *(joorkatkat)* among people and communities in the RMI so our nation and people can sustain themselves including being knowledgeable about necessary Climate Change adaptation practices.

Preventing, or at least delaying, resettlement is built into the mission of the Marshall Islands Conservation Society (MICS): by safeguarding the outer island subsistence lifestyle, the organization aims to staunch the flow of islanders from the outer islands to the urban centers and from there to the United States. Executive Director Albon Ishoda told me that he doubted adaptation efforts would be sufficient, but even so, "We're not going. We're going to stay here until it's unlivable." The newly minted NGO Jo-Jikum takes the same stance, according to all of the members I talked to. Kathy Jetnil-Kijiner, a major exponent of the organization, explained that the NGO's slogan, *Liok tūt bok* ("rooted deeply," like the aerial roots of the pandanus tree) encourages Marshallese youth to root themselves in their land like a tree, immovable. *Jo jikūm̧* itself means "your place." Jetnil-Kijiner has spoken out against resettlement in a poem performed at the Poetry

Parnassus Festival in London in 2012, and in an interview with 350.org (Jensen, 2012): "We are in no way going to just give up, pack up, and leave our islands. Our culture is rooted in our land, and our land is our life."

Here grassroots and governmental stances align. The governments of all the other coral atoll nations—Tuvalu, Kiribati, and the Maldives—have tentatively explored options for relocation (Chapman, 2012; Keane, 2009; Russell, 2009; Schmidle, 2009; Toomey, 2009): Maldives president Mohamed Nasheed made headlines in the New York Times in 2009 when he announced his interest in buying a new homeland overseas (Schmidle, 2009). But the Marshallese government has by and large rejected this sort of contingency planning. Relocation is not just *not* planned for, it is deliberately *un*planned for. President Kessai Note told me in 2007 that no evacuation should be planned or undertaken in this generation, advocating mitigation instead. A 2008 submission to the UN Human Rights Commission from the Republic of the Marshall Islands Permanent Mission to the UN stated the position clearly: "the reclassification of Marshallese as a displaced nation, or . . . 'climate refugees,' is not only undesirable, but also unacceptable as an affront to self-determination and national dignity" (RMI Permanent Mission, 2008: 10). A foreign consultant presented several government employees with a thought experiment in which sea level rise had accelerated drastically and inundation appeared nigh; even in this hypothetical eleventh-hour scenario, the officials considered the reduction of greenhouse gases to be more important than planning for resettlement (Peter Rodgers, personal communication). The government's 2006 Climate Change Strategic Plan, issued by the Office of Environmental Planning and Policy Coordination (OEPPC, 2006), contains not a word about evacuating, planning for evacuation, or procuring a postevacuation homeland. During interviews in 2007 and 2009, the official in charge of this report, OEPPC Director Yumi Crisostomo, was unequivocal in her position. In 2007 she told me:

YC: We want to live in our lands for a long time. This is for the sovereignty of the nation.

PRG: Would it be good to make a plan for possible migration, even so?

YC: The reason for promoting resilience is to *prevent* that from happening. Some people say, "Migration will have to happen eventually," but I say, "Well, take you for example. You're going to die eventually, right? But that doesn't mean you're not going to do anything now." . . . We have different views on migration from Kiribati. This is our land. . . . There are things you can do within this country. For instance, you can use customary practices to help conserve land and resources.

On the issue of climate change migration, Marshall Islanders find themselves in an awkward position in which preparation is a form of surrender and the line between determination and delusion is thin. Some would call Crisostomo the Winston Churchill of climate change, refusing to surrender her island to the foreign menace. But others would call her the Neville Chamberlin of climate change, refusing to acknowledge the gravity of the coming threat and therefore only lending it more strength. That is the relocation debate in a nutshell; but to most Marshall Islanders, she appears more like a Winston Churchill.

A few well-educated Marshall Islanders lamented, on purely pragmatic grounds, that the citizenry and the government had no Plan B for resettlement:

> We don't know how many years from now when the country will be submerged. Will it be five, or ten, or what? . . . It's scary . . . to think this place will go down under water. . . . But if it's true, then we should know how much time we have, and what preparation we need.
>
> Do we have any choice if the sea would be really rising? Where can we plant our trees, if the ground is really salty? What if there is a typhoon, are we safe to be on these islands? Especially talking about sea level rise, we will just eat fish every day, if no plants are growing. . . . To go to another land is . . . not a comfortable idea. But the question is, are we realistic, if the sea level rise comes and covers our island with sea water, and our plants are dying, corals are hot, the corals will die because of the climate change. Fish will disappear too. . . . That's why . . . the government needs to have a plan. We need to give the people options . . . so they can make a wise decision.

At the April 2009 WUTMI climate change forum, one group of participants dissented from the antirelocation consensus, saying that Marshall Islanders, like some government officials in Kiribati and Tuvalu, ought to begin the search for a new home. Locals are always confident that somewhere, somehow, a resettlement site could be found; no Marshall Islanders ever told me that they expected the international community to simply let the population drown. Some locals assume that the destination would be the United States, saying that America is the country that takes care of (*lale*) the Marshall Islands, that the Marshall Islands is still *iuṃwin pein* ("under the hand of"—controlled and protected by) America. But many, when asked where locals would flee if the islands sank, give wider answers: "America, or Australia, or Japan, or places like that"; "Whoever is willing to take us"; "Somewhere high, with mountains." In these discussions the Marshall Islands are spoken of as low (*ettā*) and large foreign countries as high (*utiej*)—geographical descriptions that potentially carry metaphoric resonance. People talk of going to "the high countries" (*laļ ko reutiej*), "where it is high" (*ijo eutiej ie*), or "where there are mountains" (*ijo ewōr toļ ie*).

In 2011, Justice Minister Brenson Wase wrote to the *Marshall Islands Journal* to argue that climate change was such a serious threat that evacuation should be considered and the government should not discourage well-educated Marshallese from making a life abroad (Journal, 2011e). Later that year, the government under President Jurelang Zedkaia coorganized, with Columbia Law School's Center for Climate Change Law, an academic conference on the legal ramifications of the depopulation or disappearance of a nation. In his keynote address, Zedkaia stated,

> I can assure you that casual discussions on relocation are an irresponsible political exercise. But it is now not so distant—relocation is also [an] undeniable threat knocking at our door, and something which could well unfold in decades to come. It is a threat that the international community is presently unprepared to address, and it can no longer be ignored. (Zedkaia, 2011)

While Wase's and Zedkaia's statements hint that the Marshallese government is beginning to warm to the idea of planning for relocation, officials remain far from regarding the move as inevitable or acceptable. Easily the most commonly voiced opinion is that wholesale exodus is such a horrific "solution" that the country must not resign itself to it, at least not yet.

When I asked Marshall Islanders what would happen if the country were to be washed off the map by sea level rise, they often replied, "We will die." Others simply said they would never depart, come hell or high water. Would locals really sooner drown than leave? Even if the statements were not meant literally, they express just what an "existential tragedy" (Mortreux & Barnett, 2009: 110) relocation would represent.

> We would die. Because everything would die. There would be no land, so all the plants and animals would die. There would be coconut trees, no pandanus trees, no breadfruit trees. Nothing would survive.

> We would die. There is only one place for Marshall Islanders. There is no other place.

> We will die. We will drown.

> We'll just stay and wait and watch. I'm not planning for climate change. I say, "I will stay. Where would I go?" I say, "I'll just stay." But I'm afraid. All the times I hear about it on the news, I'm scared.

> I hope it doesn't happen in my lifetime, because I don't want to go anywhere. I want to stay right here. That's what I hope for. I know it may be inevitable. But I'm just wishing it's not in my lifetime. Because if someone tells me, "go somewhere," where? . . . Where would I go and

> be free? And have my own land, my own house? . . . I think I am one of those who is in denial. . . . When people ask me, "If it's tomorrow, everyone needs to evacuate from this island, come next month," what would I say? "I'll stay!" [Laughs.] I'll just stay and see what happens.
>
> I know there is going to be a time when this island will be under water. . . . For me that's really hard, because I feel so bound to this place. I love it and I consider it my own. . . . If this happens in my lifetime, I'd rather die with this island than go elsewhere . . . because I feel this place is part of me and I'm part of it. It's sad for me to imagine that, but it's going to happen: in the present situation there's not much we can do. Imagine if your country was going to disappear under water. . . . There are even some Marshallese who would prefer to go to the United States. Not everyone's the same. . . . But for me, this is the place, where I'm going to die. My grandma, my great grandma, they are all buried here, so I'll be buried here too. I can't imagine living in another country for long. (UNFPA, 2009: 22)

One elderly Majuro man explained the sentiment in the following way. In the old days of open-ocean sailing, a sailor would weather a squall by partially submerging his canoe, so that the winds would not snap the mast. When the storm passed, he would bring the canoe out of the water, bail it out, and sail to shore. That, he said, was the attitude of Marshall Islanders, and their response to sea level rise would be analogous.

The most extreme go-down-with-the-ship statement that I heard was from a well-educated male government worker who declined to be identified. He stated that the Marshallese response to climate change should be a sort of autogenocide: cultural suicide for a futureless nation. In 2009, he expressed it to me as follows:

> If the IPCC scientists are correct, then we'll either have to build up our islands or stop reproducing, because we don't want there to be future generations with no homes. People talk about the option of migration sometimes, but I don't like it. I'm glad that I'll be dead by then. I see people who migrate these days, and they don't do well. Children growing up in an environment like that won't grow up to be normal people. . . . I don't want my grandchildren to go live on a reservation . . . There would be no more Marshallese people, because there won't be any more Marshall Islands. Well, that, or if it's possible to develop gills! [Laughs.]

The man's joke about growing gills points to another indication of islanders' discomfort with the idea of resettlement. As in many or perhaps all cultures, Marshallese humor disguises and dampens emotional distress: people joke about what pains them most. I have seen Marshall Islanders joke about such heavy topics as Hell and being devoured by sharks, and the experience of nuclear testing is sometimes described in what are, for an outsider, shockingly flippant terms. A man from Ujae Atoll was in stitches as he related how the people of Rongelap Atoll had irradiated themselves: they had been warned not to eat local food but just could not resist the temptation when they saw all those tasty-looking crabs and coconuts. A woman from Rongelap laughed hysterically as she explained that the 1954 H-bomb detonation (the Bravo Shot) had happened on her birthday, so it was like a birthday cake from America. A man from Likiep Atoll related his childhood memories of the Bravo Shot, producing peals of laughter in everyone listening: when the roar of the atomic blast reached the island, he was sitting in the outhouse, and he ran outside in terror without pulling up his pants—"They say that people on Likiep were exposed. Well, I was doubly exposed!"

This sort of dark comedy is, in the words of one scholar and long-time resident of Micronesia, a "playful, satiric humor with a distinctive Pacific stamp that could be drawn upon even in the darkest of days" (Hezel, 2009b: 14). It often comes out in discussions of climate change—but not at just any time, for instance when talking about whether climate change is happening or who is to blame for it, but quite specifically when imagining the very worst that climate change could bring, the total destruction of the Marshall Islands and the death or displacement of the Marshallese people. This frequently brings on not just nervous tittering but hearty laughter. Sometimes the humor comes out as a facetious way of describing this scenario: "We'll float!" "Ujae won't be very good for living, but it will great for spearfishing!" "The capital building will make a great dive site." Sometimes it comes out as a knowingly ludicrous suggestion for solving the problem: "We'll swim!" "We'll just climb up the breadfruit trees!" "We'll live in water catchments!" "Donate life preservers. Our kids and grandkids will need them." "A ship will bring a mountain from America!" "Why don't you Americans give us the Rocky Mountains to live on?" "We can go live in California—Obama will give us half the state to live in!" "Has anybody claimed Antarctica?" Sometimes there is no witticism at all, but simply laughter while describing the horrific prospect in plain terms. The following statements were all said with heavy laughter: "The islands will be gone," "It will be a barren island with no people," "Marshall Islanders will be extinct. There will be no Marshall Islands and no Marshall Islanders." Locals actually go out of their way to joke about it. In 2009 I heard a conversation between three Majuro men:

A: If climate change is happening, they have to do something about it.
B: We cannot do anything except pray to God.
C: We'll die for it. [Everyone laughs loudly.]
A: If it happens, Christian believers will say, "Everything is in God's hands!" [Everyone laughs loudly.] "Just wait and see!"

C: Another thing that climate change is causing is more typhoons and hurricanes in the big countries. So where would we go? If we go the high countries, we would die there too! [Laughs.]

C: I heard a joke. It goes "When the country sinks, will the last person to leave the Marshall Islands turn off the light?" [Everyone laughs loudly.]

B: People have to stop smoking. It's part of the greenhouse effect! [Everyone laughs loudly.]

Climate change gallows humor, I believe, reveals just how serious the topic is to people.

THE VALUE OF THE HOMELAND

Why reject relocation, even as a contingency plan? Certainly one of the most memorable exhortations to relocate—Makoni Pulu's presentation narrated above—was poorly pitched. A team of medical researchers quote the mother of a young child with cerebral palsy: "The doctor simply stated the diagnosis, handed us some papers with general information . . . told us to institutionalize him, and walked out the door. We would like to have been treated like human beings" (Sharp et al., 1992: 544). Pulu's message had all the sensitivity of the doctor's. He adopted a confrontational and blunt tone in a country where respectful circumlocution is expected when one is publicly voicing controversial opinions. Equally bad, he everywhere inadvertently trampled on the narrative of modernity the trickster. The presentation began well enough, as Pulu implied that climate change was part of the more general currents of a modernizing world. Thereafter he faltered. He failed to place blame for climate change on Marshall Islanders' overreliance on foreign things; blame was not even mentioned. Without providing an ideologically appealing reason to embrace the threat, to turn it into a *risk* in Mary Douglas's sense, he had done nothing to prevent his audience from succumbing to the optimistic bias, the simple desire to disbelieve that their country was doomed. Nor had he addressed the biblical interpretations that sanctify such disavowal: he took the reality of climate change for granted, ridiculing skeptics rather than approaching them on their own terms. Pulu in fact reversed the idea of modernity the trickster by 180 degrees. The villains of his tale were those who clung to the old, while the heroes were the ones who welcomed the new. The relocation leaders were to be the youth, the followers the elderly, in an inversion of the usual hierarchy. Pulu had blundered into framing climate change concern as untraditional, something that youth embrace and elders lament. It did not help that Pulu himself was an outsider. Polynesians are considered distant cultural cousins, but not insiders: heeding Pulu's message smacked of being taken in by the persuasive falsehoods of foreign ways.

Another reason for the reluctance to relocate is the notion that there would be nowhere on earth to hide from the ravages of global warming; migration was therefore pointless. A few locals with a particularly apocalyptic view of climate change voiced this opinion. An elderly woman from Rongelap told me, "When we run from the warming [*okmāāṇāṇ*], where will we go? Only the sky won't be hot!" A well-educated man involved in education in Majuro said:

> We can relocate to a place where we can stand for a little bit longer, let's say US. . . . [But] then that's only temporarily. It's a global warming of the earth—where can you run? Where can you go to? I mean eventually it will catch up with the US. . . . There's nowhere you can run to on earth. Unless you go to the moon or we go to Mars. [Laughs.]

One reason for declining to *plan* for migration is that there is no pressing need to do so: Marshall Islanders already have the right to migrate to the United States under the Compact of Free Association, and, as we learned above, they believe that countries other than America might also be willing to accept migrants. These safety valves rob urgency from the issue of evacuation planning.[3] But this explanation is superficial. Yumi Crisostomo dismissed the notion that the Marshallese government and people rejected evacuation planning simply because they already had an escape route. In 2009 I asked her opinion of Maldives President Mohamed Nasheed's plan to purchase land abroad, and she replied that the situation in that country was different because, unlike the Marshall Islands, the Maldives had no convenient preexisting migration corridor. I then asked if this was why the Marshallese government had made no plan for evacuation. She rejected this emphatically:

> No, no. The main reason is that these islands are given by God. Land and people are tied together. You can go live on another land, but your heritage is in your land. It's like the Kurds—they live in Iraq but it's not really their land. That's the kind of situation we want to prevent from happening to Marshallese people. We need to build resilience in order to stay in the country. We have a right. Every Marshallese person has heritage land. Everyone here can count back generations to this land. Most people in the world don't have that, but we do.

Arguably, this stance confuses two distinct meanings of "plan": planning *for* possible migration is different from planning *to* migrate, the former being a no-regrets strategy of preparation, the latter a risky gambit of preemption. But this argument is misplaced, because even planning for *possible* future evacuation implies that relocation is an acceptable response. A senior government official pointedly indicated this when I asked him in 2009 about the possibility of moving to Indonesia, a country which had recently announced that it would consider renting uninhabited islands to climate change exiles. He said:

> That is not a viable option. Once small islands agree to that, then we become dispensable—if that's the case, then you could just move us to Missouri! . . . People will not migrate. That would make them expendable, as they were in nuclear testing. Migration is *not* a solution.

Later he put his position even more resolutely: "Relocation is *not* a solution. It would be like genocide. There would be no more Marshallese people—no language, no culture. That is *not* a solution."

These opinions were voiced with more than a little emotion. Indeed, when Yumi Crisostomo pointedly remarked that "The main reason is that these islands are given by God," it seemed that by being pushed to justify the government's lack of a contingency plan, she had been forced to dispense with officialese—detached, jargon-laden talk of building resilience and strengthening partnerships—and acknowledge a more basic motivation: the idea of losing the country pained her personally, and she (and her government) were reluctant to plan for it because it was too horrific a scenario to contemplate. This is closer to what, I believe, is the more fundamental reason for the Marshallese rejection of relocation, an aversion at once visceral and philosophical. The abandonment of the country, it is often said, would kill *m̧antin m̧ajeļ*. "If they flee to America, they won't know their own custom. They'll follow what Americans do. They'll wear shorts instead of trousers." "It would be like the Bible story of the Israelites in Egypt, not being able to follow their culture." "I think that would be the end. That will be the end of the culture and the end of the Marshall Islands." "Our lives would be history, we would be studied in history." "It will be very hard, because you're living by another culture. There may come a time when these islands are submerged and sink, and so will the custom and the culture." At the WUTMI climate change forum in April 2009, two speakers—Moriana Phillip from the Environmental Protection Agency (EPA) and Yumi Crisostomo—directly stated that the disappearance of the country would doom the culture.[4]

Some locals gave somewhat more optimistic statements, saying that evacuated citizens would keep Marshallese culture in their throats (the seat of emotion in Marshallese idioms, equivalent to the heart in western usage) (*Kōjparok m̧antin m̧ajeļ ilo buroier*), acting like Americans but still retaining a Marshallese identity. Similarly, a man cited the proverb "The parrotfish does not forget its home" (*Ekmouj ejab meļọkļọk kōlñe eo an*), referring to the fact that this fish always returns to the same reef-hole (*kōlñe*), and metaphorically to the fact that people remember their origins. Others said that if the evacuees continued to live in close proximity to each other, or on an island, they could perhaps keep a semblance of *m̧antin m̧ajeļ*. A few locals likened climate change evacuation to Noah's journey on the ark, with a resulting guarded optimism that some parts of the culture, like Noah's animals, could be saved from the deluge. Locals often feel that *current* Marshallese immigrants to America have retained some elements of Marshallese custom. Although many speak of Marshallese transplants living by money, adopting American culture, the women wearing shorts, drinking, and flirting—all of the familiar laments about modernity's seduction—not all do. Some emphasize that Marshall Islanders in Arkansas, California, and so forth still give tribute to their chiefs when they visit, still gather for *Keemems*, and still *lale doon* (take care of each other). But these same optimists often say that if the islands disappeared, if the link to the homeland were severed, *ṃantin ṃajeḷ* could not survive. As a middle-aged Majuro man said:

> Let's say there's no country to write back to, to communicate with people in your home country—you've lost all of that . . . Now Marshallese people go to Arkansas and they still can pick up a telephone and talk in Marshallese to a relative here, or send a mail. . . . Without that, it'd be easy to lose the culture altogether. . . . We're basically going to be saying, we've lost our language, we've lost our identity, we've lost our country.

Evacuation is often spoken of as pulling the plug on the already ailing patient that is *ṃanit*. Resettlement strikes at all the pillars of tradition. The most fundamental loss would be land (*bwidej*). Well-educated locals, in particular, point explicitly to the cultural importance of land. "Without our lands I don't know how we can keep our culture." "Our culture is attached to our land." "For Marshallese people, you are all connected to your land. Your land is your identity. So if we go somewhere else, you don't have an identity. You've lost it." These statements point to the importance of land but do not explain it. Is this because Marshallese attachment to their land is ineffably mystical or fundamentally unintelligible to westerners? I do not believe so. The cultural value of the Marshallese homeland is quite cross-culturally understandable when local views of tradition and cultural change are taken into account.

Recall from Chapter 1 that locals say that they have always inhabited the Marshall Islands, that it is their only and eternal homeland, God-given. The value of the country is not only that it has been given to Marshall Islanders but also that it has been given *only* to Marshall Islanders. This is crucial because Marshallese believe strongly in the power of cultural imitation. Children are abandoning Marshallese ways because they are adopting the ways of the Americans, Chinese, and other foreigners who surround them. A middle-aged Majuro man said, when I asked him if Marshallese evacuees could retain their culture:

> Look at the way those kids grow up in Arkansas. . . . My sister was in the military with her husband once, back 19 or 20 years ago, and when

> they came back, they had three kids [with them]. [All] of them [the children] went back to America . . . [because] none of them could speak the Marshallese language. They cannot relate to anything Marshallese. . . . One of them grew up in Japan. He's basically Japanese. So it depends on where you grew up, and the influence. We tend to be influenced by the society and the surroundings.

I asked President Kessai Note the same question in 2007:

> It will be very difficult, because you would be living with another culture and so forth, with people from many countries, not just US. There are Orientals, African Americans, from South America, all over Europe, all over Asia, many cultures. There may come a time when the Marshall Islands is submerged and sinks, and so will the custom or the culture. Going to another community or another place, we have to be adaptable and so forth, meaning you're compromising, so you still will survive in the new environment and new culture. I hope that doesn't happen. Sadly that may be the scenario in the future.

Marshall Islanders, then, believe in their own adaptability—they will assimilate into whatever society they find themselves in—but this is not a happy belief, because it is seen to doom Marshallese culture in the case of evacuation.

The homeland is considered *unique*. This uniqueness extends from the level of the country as a whole to regions within it, to individual atolls, and to individual islets, land parcels, landmarks, and reef and ocean features. The language contains some 7,000 place names (Abo et al., 1976), an average of almost 40 per square kilometer of land. *Kapin meto*, the "bottom" or western end of the ocean, comprising Ujae, Lae, and Wotho Atolls, is renowned for its traditional navigators. Individual atolls have reputations encoded in sayings like *Namdik alele eo* ("Namdrik is the *alele* basket"), referring to Namdrik as a place full of cultural valuables (like the *alele* basket)—such as canoe-building skills, singing ability, and traditional dances—and *Aelōñin Jālwōj ej aelōñin jālele* ("Jaluit Atoll is the atoll of meat") referring to its abundant marine resources and thus a lifestyle of easy subsistence. Ujae Atoll is renowned for its traditional Jebwa dance, Ebon Atoll for its rainy fertility and for being the landing place of the first missionaries.

To the Marshall Islander, the islands are old; they have always been there, and they are filled with history. Islands are the resting places of ancient chiefs. Legends (*bwebwenato* or *inōñ*) rarely take place in generic or mythical locations. They take place on the familiar land tracts, islands, and atolls of everyday life. Small landmarks and reef features have associated legends: a rock on Mejit Island is said to be an oracle that tells boys if they will be able to grow beards; boulders on Bok Island in Ujae Atoll are said to have been thrown from Ujae Island in a fit of jealous rage by the legendary character

Joalon; small hills on Ebeju Island in Ujae Atoll are considered the site of the first performance of the Jebwa dance, and a depression in the ground is where the spirit (*ṇooniep*) who taught the first Jebwa dance is said to have appeared. Even Wake Island (*Āneenkio*), never permanently settled and now an American possession, was home to unique species of bird and flower with ritual significance (Spennemann, 1991); even uninhabited land can be cultural, an irreplaceable place. If people were somehow able to construct a new Marshall Islands with the same resources as the old—not as fanciful as it may sound, considering the Maldives government's proposal to build artificial "safe islands"—the loss would still be profound.

Failing to appreciate the uniqueness and irreplaceability of the islands is a common outsider's error. The US military made this mistake when they blithely moved the people of Bikini to Rongerik, reasoning that one coconut-fringed islet was as good as any other. Relocationists make it again when they imply that any foreign plot of land with a favorable climate and some job opportunities will make an acceptable new homeland for climate change refugees. David Parkin notes that refugees always bring with them a physical piece of their abandoned homeland (Parkin, 1999)—objects act as anchors for identity, their materiality making them durable in a chaotic world. If Marshall Islanders were forced to leave their country, they could take with them cowry shell necklaces and pandanus-leaf mats, but they could not carry the most important piece of material culture of all, the islands, or the most important human artifact, the landscape.

Land is livelihood—factually for those Marshall Islanders who still rely partially on home-grown produce, and symbolically for all Marshall Islanders. Locals give disarmingly straightforward answers to the question "Why is land important to Marshall Islanders?" It is survival, people say; that is all we have. For an outer islander, seeing your land destroyed would be like a westerner seeing his home burned down, being fired from his job, and losing his savings account all at once. Land is exceedingly scarce in the Marshall Islands, raising its value. Competition over land is not only the theme of the chiefly battles and royal assassinations of old, but also of countless disputes today: land boundaries are contested, land leases challenged, and land laws debated as a major bone of contention between the country's two political parties.

Land allows subsistence, a pillar of tradition in the local view. To lose the land to inundation or evacuation would be to trade easy abundance for *mour kōn ṃani*: "living by money," or what I have called modernity. As a Majuro man involved in marine resources management told me in 2009:

> We won't retain our original identity, where our lives are dependent on what we caught today—I go out and I fish, I make an effort and I bring back enough. Now, I [would] go to *work*. . . . I believe that living in the islands is a lot better off than packing up and taking your whole family with you to the unknown.

A poignant account of subsistence lost came from my interview with a government minister in 2007, as he imagined life after resettlement:

> We've never been exposed to dry land where there's no water around, so we don't have any skills to make a living off of the land. We can't kill snakes. We're afraid of snakes. We've never seen a snake. And there's no fish. I'm sure that when you try to kill the birds, you get citations from the people in the government, EPA or whatever, because you're not supposed to kill the birds. There's so many restrictions what you can and cannot do. When you think about it, you are almost really better just getting yourself drowned. . . . You really have to re-educate people and strip them of their cultural identity . . . your ability to fish and your ability to hunt birds. . . . You're almost like an alien on a different planet, because all of a sudden, you're placed in the middle of the desert. It might be better to wait for the waves to wash over you. . . . I'm sure some people in the Marshall Islands would prefer that . . . If you go to the outer islands, people always hunt turtles. That's part of our culture. If you do that in Hawaii, you'll be put in jail.

A young man who had attended all three of the University of the South Pacific's climate change-themed summer camps expressed his commitment to stay in terms of native Marshallese trees:

> The first year [of the summer science camp] was the first time I realized we were in big trouble in the Marshall Islands. The water will rise by seven feet. There will be water everywhere. We will have to learn to live like that. [At the summer science camp on adaptation] they told us not to leave. We are proud to be Marshallese. Adaptation is the only way to preserve our custom. . . . It's a bad thing for Marshallese people to leave the islands. In America there are no coconuts, breadfruit, or pandanus. If they leave their island, I don't know how they will teach their kids. Their kids may not know what a breadfruit is or how to say "eat pandanus"—*wōrwōr bōb*. . . . I want to go back to the outer islands after college—go to Utrik Atoll, where my mother and grandmother are from. That will be my resting place.

In 2012, when I asked Milner Okney, Public Awareness and Education Manager at MICS, what would happen if climate change proceeded apace, he replied:

> I don't think my grandchildren will be able to sail! . . . How will your children know how to make *bwiro* [preserved breadfruit] if we're migrating to the United States or even to other places around the world? . . . How can our children learn how to sail with the stars, or even fish, dive, or even weave mats? . . . The language will be there but

> language itself doesn't come with culture. *Ṃanit* . . . doesn't just come with speaking the language. You got to learn ways to survive . . . learn how to sail the stars, weave mats, make food, live in the harshest conditions that the world can give you.

Evacuation would kill land and therefore subsistence, and it would not stop there. When I asked President Kessai Note in 2007 whether climate change refugees could retain Marshallese culture, he said, "If you are in America, and you live by money [*mour kōn ṃani*] . . . it will be hard to support your sister, your father, your siblings, when your house is about to be disconnected from the electricity." An elderly man in Majuro, imagining the same exodus scenario, told me, "They will live by money, every man for himself (*kwe wōt kwe*), and not take care of each other." Subsistence lost is conviviality lost.

Land and subsistence are the foundation of Marshallese hierarchy, as discussed in Chapter 1. If the islands were abandoned to the sea, what of the traditional power structure? It could hardly survive, in the local view. To lose the land would be to lose the *reason* that low people respect high people and the *means* by which high people support low people. A government official pointed to this in 2007 when he envisioned an inundated future: "The value would be gone because of the intrusion of saltwater. . . . Why would I respect my traditional leaders? The land value's not there." Clarence Luther, mayor of Namdrik Atoll, told me in 2012 that hierarchical respect could hardly survive climate migration:

> We'll lose [the culture]. If you live with other kinds of people, you'll follow what they do and become like them. Traditional titles will be gone. Right now people respect *aḷap*s because of the land. It will be very hard to keep that. In America, at 18 years old you become totally independent. In America, you're "born free," without a title. But here, you are born, and they already say, "He'll be irooj.". . . . In America your title depends on what you *do*. You get a title by your actions, not when you're born.

Essentially, Luther's lament is about the ineluctable slide from ascribed to achieved status, and thus from traditional hierarchy to modern egalitarianism, if the material substrate of inherited privilege disappears. So Marshall Islanders tend not to trust themselves to carry on respecting higher members in the landowning hierarchy merely out of traditionalist fervor: they need the enforcement of chiefly land ownership in order to remain loyal. This is part of Marshall Islanders' belief in the allure of foreign ways—the perhaps paradoxical position that traditional Marshallese life is all-fulfilling and yet that people are easily tempted away from it.

For Marshall Islanders, *climate change is a modernity maker*. In the event of evacuation, the migrants might find themselves far from poor:

considering the media attention that low-lying island countries are now receiving, there might be a true deluge of international charity were they to actually be destroyed. The exiles would not lack for money, but they would lack for their land; it would be the ultimate land-for-money swap, not a fair trade, and exactly the thing that Marshall Islanders have been steadfastly avoiding during almost two centuries of western influence, as I discussed in Chapter 1. The forced selling of a homeland would be all too familiar for the country's nuclear-affected communities; the Marshallese would become a nation of Bikinians, on a reservation like Kili Island. A Bikinian elder famously said, "We've learned to dry our tears of sorrow with dollar bills. But money never takes the place of Bikini." If I am correct that for Marshall Islanders land is a metonym for tradition and money for modernity, the elder's message is more general: modernity dries our tears but does not staunch them. Were the country to be abandoned to climate change, all Marshall Islanders might say, "We've learned to dry our tears of sorrow with dollar bills. But money never takes the place of the Marshall Islands." So Marshall Islanders' unwillingness to resign themselves to climate change evacuation is the same as their unwillingness to sell their land; it is no coincidence that Marshall Islanders are among the Pacific Islanders most opposed to both.[5]

A skeptic might say that despite all of this, Marshall Islanders' *behavior* shows that they are perfectly willing to forsake their heritage land—and thus "tradition" in general. They are happy to migrate, in droves, from the outer islands to the urban centers, and even to the United States, voluntarily adopting the market lifestyle. There is some truth to this. Marshall Islanders *have* compromised on their attachment to land, in myriad ways. Even those who remain on rural islands choose to rely more heavily on imported food than on local produce (Rudiak-Gould, 2009a: 35–37). But Marshall Islanders have never made the ultimate sacrifice. They do not *all* move to the urban centers or to the United States; and even when an islet has been depopulated, the names and owners of all of its land tracts are remembered—and they will never be sold. Marshall Islanders venture all the way to the point of no return, but they do not cross it. That bright line—expropriation or destruction—is the point at which it would never be possible, even in one's imagination, to return to *ṃantin ṃajeḷ*. What appears to be a small step—selling off a land tract that is already uninhabited and providing for no one—is in fact a giant leap. When Marshall Islanders leave their heritage land, they are giving up tradition, but not the idea that one day, perhaps, they can return to it. To lose that land entirely would be to surrender forever the hope of returning to a state of grace: not only exiled from the garden but also barred from heaven. As one government worker said, imagining climatic exodus, "We [would] be a different kind of refugees, running from climate change. Other refugees, they have a place to go back to. But if you run away from the rising sea, you can't go back."

So climate change exodus is no plot twist in Marshallese history: it is, in local perceptions, the final act in a two-hundred-year tragedy. Even if it is inevitable, to succumb to it—however reluctantly, however provisionally—would be modernity the trickster's ultimate swindle.

ADAPTATION AND THE AFFIRMATION OF TRADITION

Marshall Islanders have chosen instead to pursue in-country adaptation. The English word *adaptation* now sometimes appears in the Marshallese language, and WUTMI has recommended translating *resilience* as *joorkatkat* ("Stand ready; mobilize (military); fighting stance" [Abo et al., 1976: 116]), probably related to the word *joor* ("pillar; column") and *joortoklik* ("security; land or goods or money put away for future use or for children; insurance; . . . guarantee; . . . savings" [Abo et al., 1976: 116])—all of which convey stalwart strength against the forces of decline.

Most of the projects have entered the scene since 2009. The Ratak Atoll mayors' request for US$51 million from the UN includes a request for reverse osmosis water making units to relieve climate change-exacerbated water shortages. In September 2010, a team—including Prof. Paul Kench (a University of Auckland coastal geomorphologist), Dr. Murray Ford (a marine researcher based at the College of the Marshall Islands), and representatives from MICS and the EPA—partnered with the outer island communities of Jeh Island (Ailinglaplap Atoll) and Jabat Island; their plan was to use surveying methods in order to determine how high above sea level their homes, crops, and other assets lay, with an eye toward future adaptation to sea level rise (Journal, 2010j; Kench et al., 2011). Another project with a high-tech component was an initiative in early 2010 by Arno's Kobamaron ("power together") community organization, financed by Germany's Federal Foreign Office Task Force for Humanitarian Aid and assisted technically by coral reef experts from Pacific Aquacultures Cooperatives and the Global Coral Reef Alliance. The project sought to test the feasibility of Biorock technology: electrified wire meshes, powered by wind, wave, and solar power, were installed in the near-shore lagoon waters of two Arno Atoll communities and a heavily visited outer islet of Majuro Atoll. Preliminary results indicated that, as anticipated, the electrified wire structure promoted rapid coral growth, raising hopes that a larger-scale deployment of this technology could strengthen reefs and thus prevent the synergistic climate change impacts of shoreline erosion and coral death (Jormelu, Hagberg, & Goreau, 2010). Less futuristic are the seawalls built of concrete, piled rocks, oil drums, and any other available materials that Majuro families have constructed to protect their shoreline homes. A community action group on Majuro named Rita Reimaanlok has recently been awarded funding from the Australian government to install concrete blocks in the Majuro lagoon to deflect wave surges as well as provide sporting and recreational opportunities.

These "untraditional" and high-tech adaptation strategies are the exception: most Marshallese adaptation efforts are founded on techniques considered traditional, which ordinary Marshall Islanders can implement without outside funding or assistance. WUTMI's 2009 climate change forums encouraged participants to conserve freshwater resources: water shortages already pose a serious problem, yet one that is manageable in the short to medium term and is familiar to a society that has inhabited, for millennia, drought-prone islands with very few lakes and no streams. A community festival in Rita in October 2011 (the grassroots portion of an ADB-funded, OEPPC-organized community vulnerability assessment) was aimed at raising awareness of climate change adaptation, in particular in relation to water resources. The University of the South Pacific's Majuro campus devoted one of its climate change-themed summer science camps, in 2011, specifically to climate change adaptation. Participating high school students were taken on field trips to Majuro's water reservoirs, its coral reefs, and areas of erosion on the shoreline; the students built miniaturized demonstrations of adaptation techniques such as placing houses on stilts.

A major push in grassroots adaptation activity has been coastal management, in particular replanting the shoreline with native tree and bush species. This measure has been touted in WUTMI forums, by MICS staff, in a report by influential Marshallese educators on Education for Sustainable Development (Heine, Maddison, & Rechebei, 2009), and in the Jañn̄ōr Windward Forest Project (described below) as an effective means to simultaneously counter erosion, strengthen food security, and revitalize culture. At a WUTMI-organized pandanus festival and parade in April 2010—which celebrated the fruit as a boon to tradition, nutrition, and sustainability—the Majuro Coop School float praised pandanus trees as protectors against climate change-induced coastal erosion. Some antierosion techniques have been piloted through the Bikirin Island Conservation and Restoration Demonstration Project, an initiative funded by the Global Environment Facility and implemented by those claiming traditional rights (who have established a kin-based organization called ELEFA[6]) to the small outer island of Bikirin in Majuro Atoll. The goal is to transform the islet into an exemplar of good environmental management. In addition to ELEFA's work in safeguarding reefs and restoring habitat for the endangered native *m̧ūļe* bird (*Ducula ratakensis*), they have shown a keen interest in climate change and now use the islet as a laboratory to test erosion-fighting techniques: planting coconut and pandanus trees along the shoreline; felling inland coconut trees in the hopes that their decomposition will raise the elevation of the atoll; and surrounding nearly the entire shoreline with palm fronds stacked in piles parallel to the shore, mixed with coconut husks, to act as a seawall. They point to many virtues of this sort of seawall: it is considered traditional, uses readily available local materials, and blocks waves while still allowing those waves to naturally build up the island by washing sand and coral rubble onto the shore through the cracks in the palm fronds. ELEFA members pay regular visits to the island to report on erosion, changes in terrestrial and marine ecology, and the success or failure of their adaptation experiments.

The Bikirin project has also helped to educate the younger generation whose future depends on these adaptation efforts. Majuro's Assumption School, with the participation of American students and teachers from Honolulu's Maryknoll School, focused their 2012 summer program on climate change. Assumption ninth graders attended science classes with special attention to global warming, participated in a beach cleanup, and stayed overnight on Bikirin. Field trip organizers, including the climate change researcher and activist Mark Stege, showed students the adaptation measures that ELEFA was piloting and taught students which tree and shrub species strengthened the shoreline and which ones weakened it. The "good" trees mentioned were all native and were referred to using their Marshallese names: the *kotōl* (*Terminalia catappa*), *kōṇṇat* (*Scaevola taccada*), and *kōñe* (*Pemphis acidula*), all salt-resistant trees and shrubs that buffer against the ocean's salt spray. In the "bad" category was the casuarina tree, which Mark Stege told students was a foreign import, worse than useless for shoreline protection because its roots grow out rather than down, thus dislodging great quantities of soil when they fall—potentially a neat metaphor for the adaptation benefits of tradition though not necessarily

Figure 5.1 Assumption School students on Bikirin Island plant a *kotōl* tree to form part of a *jaññōr* forest. In the background, a "traditional" palm-frond seawall can be seen (Photo by the author).

intended in that allegorical way. The students planted several *kotōl* trees near the shoreline (Figure 5.1).

Local climate change researcher and activist Mark Stege has spearheaded the Jañn̄ōr Windward Forest Project, funded by UNESCO and the Secretariat of the Pacific Regional Environment Programme. Its origins were in a hands-on adaptation demonstration conducted as part of the Ministry of Education's climate change-themed Education Week in 2009, described in Chapter 3: students at Marshall Islands High School planted native tree seedlings on the windward shoreline and then constructed windbreaks (*jālitak*) out of coconut fronds to protect them. The project has now evolved into a science curriculum and laboratory development initiative centered on Marshall Islands High School, involving teachers, students, and nearby community members from Rita and Aenkan electoral wards. Activities include more extensive tree plantings, topographic surveys of the high school's campus and surrounding community to predict the local impacts of sea level rise, and curriculum trials that teach climate change adaptation and mitigation through an integration of material in earth science and Marshallese language arts. The campus is to be used as a living laboratory for student-led experiments on climate change impacts. The goal of the initiative is not only adaptation but also the "climatization" of the high school curriculum by merging lessons in science with lessons in Marshallese traditional culture.

At the heart of this project is its namesake, the *jañn̄ōr*, and the related word *jālitak*. There is some disagreement as to the meaning of these two words. The Marshallese-English Dictionary defines *jañn̄ōr* as "windbreak" (Abo et al., 1976: 85) and *jālitak* as "bulwark, defense, protection" (Abo et al., 1976: 89). Some climate change activists say they are synonyms or dialectical variants with the same meaning, while others say that they are distinct terms, with *jañn̄ōr* referring specifically to the windward shoreline forest, *jālitak* to windbreaks more generally, or *jañn̄ōr* referring to the outermost row of trees, *jālitak* to the row that is further inland. Whatever the exact meaning, the two words are now being used, often interchangeably, by climate change adaptation activists to refer to a windward shoreline forest, considered traditional, that acts to protect inland trees and crops from wind damage, block salt spray from contaminating the soil and the freshwater lens, and bind the coastal soil together to reduce erosion—all while providing useful tree products. It is typically a mixed forest of several species, including the *bōb* (pandanus, *Pandanus carolinianus*, *Pandanus fischerianus*), *ni* (coconut, *Cocos nucifera*), *lukweej* (*Calophyllum inophyllum*), *kiden* (*Messerschmidia argentea*), *kōn̄n̄at* (*Scaevola taccada*), *utilomar* (*Guettardia speciosa*), *kōn̄o* (*Cordia subcordata*), *pin̄pin̄* (*Hernandia nymphaeifolia*), and *kōn̄e* (*Pemphis acidula*) (Spennemann, 1993: 120–136).[7] Key in this forest is the pandanus tree: provider of fruit eaten fresh and preserved, weaving materials, and more, the tree to which an entire WUTMI festival was devoted (Journal, 2010c) and the species so often on the shortlist of traditional Marshallese crops, its monosyllabic Marshallese name (*bōb*) being rattled off along with those of the coconut (*ni*) and breadfruit (*mā*) when islanders wax romantic about Marshallese subsistence. Mark Stege has returned to his childhood home and heritage land, Maloelap Atoll, on a mission to collect pandanus seedlings to plant in Majuro as part of the Jañn̄ōr project—the traditional peripheries recolonizing the modernized center. In 2012, Marshall Islands High School students and community members planted these seedlings along the campus's shoreline. Rimel Daniel, a recent Marshall Islands High School graduate who won the Jañn̄ōr Award for being the most promising student to become a science teacher, described the pandanus planting exercise as beneficial not only for blocking salt spray but also for providing weaving materials, food, and medicine. The earlier disappearance of this sort of forest, not only from local shores but from local minds, he attributed to the general Americanization of the islands.

Perhaps the organization that has done the most to promote climate change adaptation in recent years is MICS. As I described in Chapter 1, the NGO's strategy is to integrate climate change adaptation with the more general safeguarding of subsistence livelihoods, primarily in the outer islands. Marine and terrestrial conservation, coastal management, water management, biodiversity, and food security are pursued in tandem to rescue subsistence from the myriad threats of climate change, population growth, and modernization. Quite explicitly, this is conservationism for the benefit of human well-being and cultural vigor, rather than for biodiversity, endangered species, or "nature" for its own sake. MICS has in particular promoted the adaptation measures of building houses on stilts, restoring mangrove and *jañn̄ōr* forests, and relocating wells farther from the shore and in areas where there was once a taro patch (that is to say, the thickest part of the island's Ghyben-Herzberg water lens [Spennemann, 1993: 121]).

MICS also assists communities in drafting marine and terrestrial conservation plans, as part of the multination Micronesia Challenge initiative to effectively protect 20% of terrestrial areas and 30% of near-shore marine areas by 2020. Toward this end MICS staff have been passionately promoting the revitalization of traditional *mọ* areas. Reminiscent of various other Pacific societies (Beaglehole, 1941: 64; Sahlins, 1958: 7), a Marshallese chief may declare particularly resource-rich atolls, islands, and reefs to be *mọ* (taboo, forbidden), belonging to himself alone, with no subordinate tenants, unexploitable except with his express permission (Mason, 1987). Such a practice is now enthusiastically cited as evidence that conservationism is traditional. MICS officials (as well as other well-educated Marshall Islanders involved in climate change and resource management) now speak of *mọ* areas as "traditional conservation sites," "an act of sustainability by our ancestors," and "a way to sustain livelihoods on these islands, so you don't overharvest your resources." As one well-educated, middle-aged Majuro resident involved in education told me in 2009:

> In the past, they had the *mo̧* system—the *mo̧* of the chief. You needed the chief's permission to go there. And now people say, "We need to set aside certain places to conserve natural resources." But we Marshallese have already been doing that—it just wasn't in a textbook yet!

Once vigorous, this system was weakened by Japanese administrators who declared *mo̧* reefs to be accessible to everyone (Tobin, 1952: 11) as part of their bid to limit the authority of chiefs. The practice is now weak or defunct on many atolls, although a renaissance is quietly beginning with the encouragement of MICS.

One particularly visible MICS-facilitated project is the work of the Namdrik Atoll Local Resources Committee. Formed in 2008, the committee had, by 2012, secured not only national but indeed global fame by winning the United Nations Development Programme's Equator Prize for community-based conservation activity. The committee includes representatives from the atoll's fisherman's group, women's group, youth group, *aļap*s, chiefs, and the church. In a series of MICS-led consultations, Namdrik residents conducted resource mapping and reported worrisome changes that they had observed: accelerated coastal erosion, sea level rise, saltwater intrusion into wells, insufficient water resources during droughts, declining fish populations, and coral bleaching—an impact to which Namdrik Atoll is unusually vulnerable due to its shallow lagoon. The committee has planted mangroves, placed the existing mangrove forest under protection with the help of the Ramsar Small Grants Fund, and set aside a 35-hectare marine protected area through a partnership with the US-based NGO Seacology. The committee is also considering raising taro patches by a foot in order to keep pace with sea level rise and the resulting saltwater contamination of the freshwater lens. Integrated into these efforts are local income-generating and money-saving schemes including reactivating the atoll's pearl farm and installing solar energy.

Clarence Luther, mayor of the atoll and a major player in the project, described his motivations when I interviewed him in 2012. Climate change, he told me, is central to the project. He had first heard of climate change, the melting of the ice and the rise of sea levels, during his education in the 1980s and 1990s, and said that he found it plausible from the beginning. But it was not until recently, 2010, that he became frightened of what climate change might do to his community. By that year, he could see clear impacts on Namdrik Atoll: king tides struck every two years instead of only during typhoons; the shoreline had retreated in some areas by 5 or 10 feet; coastal coconut trees were falling; overall rainfall had declined and rain was unusually heavy when it did come; and coral bleaching, previously unknown, was now visible. The reason for these changes, he was certain, was global warming. Marshallese people bore part of the blame for causing it, as they used plastic and batteries and drove cars even when they did not need to. Severe impacts, he believed, would be felt within fifty years.

Low-lying Namdrik would be 95% gone, he feared. "We sleep and worry about climate change," he confessed. "Sometimes we get tired of thinking about it because it's frightening." But he was excited about the possibility of adaptation. At the Rio+20 Conference in 2012, he heard about Biorock technology and was impressed. Raising taro patches and planting coastal trees were worthwhile adaptation measures too. The raison d'être of the Local Resources Committee, he said, was the well-being of the next generation of Marshall Islanders, not that of animals and plants. Coral should be protected not for its own value but because fish depend on it and people depend on fish. Ultimately, he said, he had no good answer to the question of sea level rise. The measures being put in place only amounted to 2% or 3% of what would be enough to save the atoll—yet he felt they needed to continue trying even so. But, Luther noted, the problems on Namdrik were not from climate change alone. They also stemmed from local behavior. "Traditional foods are gone," he lamented.

> We should go back to the old style. It used to be six months between field trips, but it was not a problem. No one was hungry. Now if the field trip ship is late by a week, people get hungry! Meanwhile breadfruits are falling and rotting. The problem isn't climate change, it's *us*. If you have [imported] rice in one hand, a [homegrown] breadfruit in the other, people will choose the rice! It's a change of lifestyle. And this is only in a small way an effect of climate change, just a little bit—mostly it's *us*.

People had stopped respecting traditional *mọ* areas and the conservationist edicts of *aḷap*s because they had ceased to care for one another. If the predictions of sea level rise came true, he worried that entire land tracts would disappear, people would be forced to leave, and the end would be close for an already declining culture.

Whether the measures discussed above can actually save the country from future uninhabitability is unclear. Many physical scientists consider it dubious (Barnett & Adger, 2003; Journal, 2009c), while some social scientists argue that if local adaptation is coupled with global mitigation, low-lying nations may yet be saved (Barnett & Campbell, 2010). But for Marshallese advocates of adaptation, the worthwhileness of these measures does not hinge on that debate. For Marshall Islanders, "adaptation" is not the only or even the primary goal—as is the case in many "climate change adaptation" efforts (Berrang-Ford, Ford, & Paterson, 2011). A tradition-affirming ulterior motive is evident. The explicitly stated objectives of the climate change-focused sections of the Education for Sustainable Development document (Heine et al., 2009) include seeing "cultural and traditional practices revived" and "renew[ing] Marshallese values." The ELEFA family organization that heads the Bikirin project has described their mission as the revival of ancestral knowledge of land care. In the Jañŏr Windward

Forest Project, the mixed shoreline forest was presented as a traditional practice, blessed with a Marshallese name; therefore the event could be counted a success even if erosion continued apace. Indeed, the goal to "promote cost effective and locally available climate change mitigation and adaptation measures" was only one of four stated objectives of this project: another was to "support existing efforts to promote indigenous vegetation use" (through the "provision of food resources in the form of planted edible Pandanus"), promoting the practice of subsistence so central to Marshallese conceptions of tradition. Mark Stege has begun to use the roots of these trees as a metaphor for the value of cultural preservation: the pandanus, an important tree in the *jaññōr* forest, is, like *m̧anit,* a "deep-rooted tree with nourishing fruit, and . . . part of our limited arsenal to preserve both culture and coastline" (Stege, 2012).

MICS presents as traditional most of the adaptation measures it advocates: stilt houses, palm-frond seawalls, terrestrial and marine conservation, and the *mọ* system. In 2009, when I interviewed Albon Ishoda (at the time an integrated marine resource manager at MIMRA, later to become executive director of MICS), his traditionalist view of shoreline protection was obvious:

> Traditionally, our ancestors understood this—that this system was here for a reason. And they had their systems of management as well. But with growing economic interest and rapidly changed society, a lot of outside influence [and] more focus on making more money, [people are] chopping down the trees. . . . Traditionally, these weren't practices—cutting down trees on the shoreline. They were known windbreakers. We have traditional names for them. . . . You would rarely see people chop them down a long time ago, because that's what they were there for. Our ancestors understand their purposes.

I then asked, "Thinking long-term about all these issues, and climate change especially, do you think that Marshallese people will be able to stay in this country in the future, in fifty years, a hundred years?" Suddenly Ishoda grew contemplative and his speech became quieter; it seemed the subject was now close to his heart. It became clear that, for him, the ultimate purpose of his job was to preserve what he could of a former lifestyle of easy subsistence. He answered:

> That's a good question. That's one of the reasons that I'm really passionate about what I do. I'm trying to ensure that—and this is more of a personal answer now—trying to ensure that my kids and my future are still able to enjoy what I'm enjoying now. God forbid that we all move to Arkansas. We always joke about that. [Laughs.] That would really be losing who I am, losing my identity, just become an Arkansan. That's one of the reasons that I'm really trying to work with these

> communities to help them develop this. [We need] at least to slow down the impact—local solutions not requiring high-tech raising [of] the atoll, but living the way we used to live.

Interviewing Ishoda again in 2012 in his role as executive director of MICS, he described how he had reinterpreted technical and academic terms such as *vulnerability* and *adaptive capacity* in terms of Marshallese traditional knowledge. "Traditionally, our ancestors' adaptive capacity was very high," he argued, naming a slew of customary practices as examples: preferential settlement on the lagoon side of an island and the windward end of an atoll; a stable population; the importance of the extended rather than the nuclear family and the communal rather than the individualist ideal; the opportunities for children to learn survival skills by interacting with a broad swathe of the community rather than just their nuclear family. All of this, he said, had faded, creating the new climate change vulnerability. In communicating climate change to outer island communities, Ishoda said that it was necessary to discuss changes not only in the global atmosphere but also in local behavior. As in mitigation efforts, in-group blame crept in:

> Sometimes I think we blame too much on others. . . . We should first and foremost [think about] what we're doing. If we're contaminating the lagoon, we should take responsibility. . . . The foreign, we cannot control. But with the locals, if we say, okay let's do sustainable land management, let's do proper planning, proper resource management, proper education and awareness, community awareness, then we can raise the adaptive capacity, reducing their vulnerability in a lot of ways. But if we just say, "Well, we're not going to plant the pandanus because we're waiting on the shipment of rice, or not going to plant the breadfruit because it's of no use to us anymore." . . . For instance the pandanus, we regard that as a shoreline stabilization . . . plant. When we grow it on the beach it's able to retain some of the shoreline. But if we say, "Well we don't need pandanus anymore, we're not going to grow it," then we're going to end up with the barren land just waiting for USDA rice to come every three months. [Laughs.] . . . There's so much dependence on outside, that it's ridiculous. You go to Ujae [an outer island with a reputation for traditionalness], you don't see hungry people. They're always taken care of. They know what to do. Their adaptive capacity is strong enough that they can easily go out and bring in three fish to feed their family. But now because of so much other influences, you know especially the other influence [he makes a hand gesture indicating "money"], "no no I'm not going to do it, I'm going to wait until the money arrives." . . . [In the past] we weren't so much dependent on these things. And we were fine. We knew how much sea level was, we had an idea how much erosion was taking place, and we knew what course to take. . . . But now part of that is gone. . . . It's the

social behavior that is really the main thing that we have to learn to address when we're dealing with climate change.

Ishoda added that he would not rule out high-tech adaptation measures, but he found the government's rush to rely on such measures distasteful. The government's recently mooted prospect of building a 5-kilometer-long seawall in Majuro at a cost of tens of millions of dollars he considered "insulting," a gesture of giving up on traditional knowledge.

When I interviewed MICS Public Awareness and Education Manager Milner Okney in 2012, he expressed similar views: He considered the modern dependence on outside things "discouraging and embarrassing" for a people who had survived for two millennia without outside help in one of the world's most unforgiving environments. While he said he would accept nontraditional adaptation means, such as reverse osmosis water-making machines, if they could help save the country, he felt that traditional revival had much to offer: "*Jālitak*, *jañn̄ōr*. Coastal buffering systems? We already had that! Building homes on stilts? We did that! That's traditional hut [building]. Traditional seawalls [also]." A similar viewpoint is voiced by many Marshall Islanders with connections to islands abandoned due to nuclear testing: they attribute erosion on those islands to the lack of human caretaking. The land suffers without its people. Elderly men from Rongelap Atoll, Marshall Islands, reportedly regard the last few decades' erosion of their islands to be the result of a lack of human maintenance, during the nuclear-induced evacuation of the atoll, and of traditional practices of land expansion (Bridges & McClatchey, 2009: 145). As in the case of mitigation, government efforts are largely in alignment with grassroots efforts, although with more interest in high-tech solutions and less in traditional revitalization.[8]

Are these measures truly "traditional conservation practices"? For instance, was the *mo̧* system a deliberate attempt at conservation? Expert opinions differ: some claim that the practice was intended to safeguard resources such as the nesting grounds for sea turtles and birds (Mason, 1987: 13; also see Bridges and McClatchey, 2009: 144; Fosberg, 1955: 14), while others claim that it was merely an expression of the chief's power to claim the richest islands and reefs as his own (Milton Zackios, personal communication; Tobin, 1952: 11–12). Did the Marshallese ancestors deliberately plant or maintain *jañn̄ōr* forests to reduce erosion, wind damage, and salinization, or was this simply the forest that naturally grew along the shoreline? Were palm-frond seawalls a widespread practice in the precontact Marshall Islands? An outsider might speculate that *jañn̄ōr* forests and palm-frond seawalls are "invented traditions" and that their characterization as traditional adaptive strategies attests more to the present-day need to justify initiatives in terms of *m̧anit* and to romanticize the past than it does to an ancient Marshallese ethos of ecological wisdom. Ultimately, the debate is academic. The *effect* of practices like the *mo̧* constitute traditional ecological knowledge in itself (see Rappaport, 1979), rendering the question of deliberateness moot.

More to the point, whether an ethnohistorian would consider palm-frond seawalls traditional or *jaññōr* windbreaks deliberate matters little. They are being cast as such now, and that is what matters.

The best evidence that Marshallese adaptation efforts are more fundamentally about modernity the trickster than about climate change per se comes not from these intimations of a traditionalist motive but from people's unoptimistic estimations of the ultimate feasibility of adaptation. Even as they speak to inspire adaptation, activists (such as the spokespeople of WUTMI) never state that such measures will save the country. One instead hears many statements to the contrary, even by the staunchest advocates of resilience building. Namdrik Mayor Clarence Luther, instrumental in the country's most celebrated adaptation project, told me that the measures he was helping to put in place would solve only 2% or 3% of the problem. In my interviews with three key MICS staff—Albon Ishoda, Milner Okney, and Carlton Abon—all emphatically said no when I asked if adaptation would work. Why pursue adaptation if it is doomed? Albon Ishoda described his motivations:

> We're going to stay here until it's unlivable . . . So if we're going to stay here, [we need to ask] what are we doing ourselves to make [us] stay, make us more resilient? . . . It's better to go fighting than just to sit. . . . We weren't a society that just sat around and let others take care of us. . . . If we were able to live on these islands for what, two thousand plus years, why are we complaining now that we can't? . . . We never know what the future holds for us, but at least when I leave I know I have left fighting. . . . Not just coming to become a dependent on, becoming a burden on another society. . . . By going to Arkansas you become dependent on welfare. But that's not what their ancestors were about. That's not who they were. They were self-dependent.

Mark Stege felt that adaptation was both futile and worthwhile. He once told me:

> What does "climate change adaptation" mean? Adaptation is just a word. The truth is that people are going to die—in floods, heat waves. The way I see it, when we're doing tree-planting as a climate change project, we're fortifying not our shoreline, but our culture. So if you can use traditional medicinal trees, for instance, to fortify the shoreline, then let's fortify the shore. If climate change causes us to strengthen our culture, that would be a good thing.

In an article published online, he wrote:

> It may be too late for the people of Taroa [his home island] to hold back inundation, storm surges, erosion and other natural processes

> that are now amplified by climate change. But for now, the way I see it, climate change for Taroa will boil down to ensuring healthier coastlines, more sustainable methods of existence, and preparing our children to articulate their own imperiled future on or off these fragile islands. (Stege, 2012)

At other times he conveyed a similar message using biblical narratives with traditionalist overtones:

> When I discussed . . . the option of focusing on cultural preservation as an adaptation strategy, this seemed to give people a welcomed alternative. I likened it to Noah preparing his cargo two-by-two and the Jews preparing to undergo a long journey before finding the land of milk and honey. . . . It meshed with something I heard for the first time from Honorable Ninwoj Lakjohn, Mayor for Wotje Atoll Local Government. . . . Lakjohn likened [climate change prediction] to the biblical warning given to Noah prior to the Great Flood, in that Marshallese now have the benefit of a warning that we must heed in order to do the right thing and prepare for inevitable sea level rise.
>
> Think of the Alele museum [a repository of traditional Marshallese material culture] as a Noah's Ark. Two by two, we take it all with us. I think that climate change adaptation in the Marshall Islands includes strengthening the Alele.

For this activist, devastating sea level rise is as inevitable for Marshall Islanders as it was for Noah, yet locals must nonetheless work to save what they can from the deluge.

Thus adaptation, like mitigation, is considered both possibly futile and definitely worthwhile. Climate change is used rather than solved. It is only through understanding the threat's framing in terms of modernity the trickster that one can resolve this paradox of action without empowerment, and acknowledge that the Marshallese insistence on adaptation before migration is neither sincere delusion nor insincere rhetoric, but a futile mission in service of an achievable one.

Conclusion
Making Sense of Climate Change

The people of the Marshall Islands have been told their country is doomed and have seen confirmatory omens in their backyards. This book has endeavored to show how locals react to this specter and why. It has shown how the idea of global anthropogenic climate change, originally a foreign scientific notion, has become local, disseminated through Marshallese society and interpreted through Marshallese concepts and values. Marshall Islanders have, to use Broad and Orlove's (2007) coinage, "channeled globality," remaking a global discourse in their image and harnessing it for their own parochial ends. Yet even as the Marshallese vision of climate change is culturally specific, it is a recognizable instance of that larger discourse, a variation on a theme. Marshallese views of climate change are in fact *part* of the global discourse, and the existence of such local instances in a variety of societies is what makes the discourse global. Thus Marshall Islanders, through climate change, have become "transnational locals" (Lahsen, 2004).

Marshallese appropriations of climate science are best understood as reinterpretations rather than mistranslations. To be sure, some Marshallese views of climate change are empirically iffy: the 2009 solar eclipse was not in fact causally related to the burning of fossil fuels and the release of greenhouse gases. But many local understandings could be embraced, or at least respected, by even the most hard-nosed physical scientist. I argued in the Introduction that climate change is not merely an environmental issue, not only a geophysical phenomenon, but also a cultural issue and a social phenomenon. If this is a point that some physical scientists and westerners need to keep relearning, it is not so for Marshall Islanders, who never thought of global warming as "environmental" in the first place. Their interpretation of climate change as cultural decline and seductive modernity makes the phenomenon social from the beginning, and it would be hard to argue that this is a misconception. It is an *insight*, and one that may help us tackle the problem in a more enlightened manner (see Crate, 2008; Leduc, 2011).

The localization of global warming discourse has taken place according to the general cultural theory of risk. Most aspects of the scientific discourse of global warming—in particular its degradationalist narrative, its anthropogenic causation, and the western origin of its culpable technologies—are perfectly compatible with prior Marshallese commitments. Those aspects are taken up and embraced. One aspect is contrary to prior beliefs or even unintelligible in terms of them: the presupposition that climate change is essentially a phenomenon in "nature." This aspect is rejected or ignored. Another aspect of the science is *ambiguous* with regards to local beliefs: climate science offers grounds on which to blame the in-group as well as grounds on which to blame an out-group. This ambiguous aspect is made to conform to preexisting Marshallese tendencies by downplaying the ideologically inconvenient proposition and emphasizing the more palatable one. The idea has been transformed in the process of its transmission; the message has been reconstructed as *traditionalist*, *anthropocentric*, and *local*, framings that the original formulators of global warming science neither intended nor anticipated.

Thus, rather than adapting to climate change, Marshall Islanders are adapting climate change to themselves. There are, to be sure, potential drawbacks in the Marshallese appropriation of climate change as seductive modernity. With the in-group blame that results, locals may needlessly take on guilty feelings for a tragedy they have hardly contributed to and lose their opportunity to lodge a powerful and moving complaint against the inaction of industrialized nations on climate change. Rejecting evacuation, they may fail to prepare for the worst-case scenario and suffer even more greatly as a result. Emphasizing tradition as the answer to climate change, they may ignore effective high-tech remedies. But if the dangers of the Marshallese interpretation are real, the opportunities are equally so. We should not endorse Marshallese interpretations of climate change on merely relativist grounds, but neither should we too hastily condemn the strategy for its various potential faults, its failure to see beyond certain parochial implications, for it is this framing that renders the threat locally meaningful and actionable.

We see, then, a sort of climate change resilience other than the physical and social varieties: *ideological* resilience, and of an impressively strong kind. Even climate change—a momentous idea if there ever was one, ideally placed, it would seem, to alter people's perceptions of themselves, their nation, and their future (Jasanoff, 2010)—has effected little or no change in local ideology. It has only strengthened it. In Al Gore's famous phrase (Gore, 2006), climate change is an inconvenient truth; for Marshall Islanders, it is a convenient one. While global warming poses a severe threat to Marshall Islanders' homes and livelihoods, it poses no threat to their concepts. It insults their islands, but flatters their categories. It is a material upset, an ideational windfall.

In this way, the case study points to a more general conclusion about risk perception, a proposition latent but never explicit in Mary Douglas's writings on risk. People are not risk-averse per se. They do not dislike risks. In fact, they like risks, they love dangers—as long as those risks and dangers flatter their worldview. Why else would most Marshall Islanders choose to believe in—indeed, in some sense to welcome—such a horrific idea, when resources exist to deny it? Confirmation bias has trumped optimistic bias. And why else, moreover, would Marshall Islanders assume responsibility for a colossal crisis, when they have every justification necessary to point the finger at others, to wash their hands of the entire affair? Ideological convenience has trumped practical convenience. Any instance of confirmation bias demonstrates how motivated people are to defend deeply held beliefs, but the sort of confirmation bias at work in the cultural theory of risk demonstrates the point especially well, because it shows that this motivation may indeed be greater than the motivation to believe that one is invulnerable. At the very least, in a situation where the risk is largely future-oriented, not yet overwhelming, the most significant dangers and opportunities are conceptual rather than physical. A threat may one day cause social upheaval through its physical impacts, but the more immediate danger is that it will cause social upheaval through its conceptual impacts. That, the ideological hazard, is the first danger, perhaps the primary one, that a society must neutralize when it becomes aware of a threat.

When a society has thus managed to ideologically domesticate and "house train" the threat, it has entered a paradoxical state vis-à-vis this threat, a paradoxical state at the heart of risk perception. Members of the society simultaneously loathe and love the threat, and both for the same reason. They loathe the threat because it endangers something fragile and dear. They love the threat because it endangers something fragile and dear, and thus societal attention on that threat will uphold the ideological scheme in which that thing is considered fragile and dear. In the Marshall Islands, if climate change is a kind of cultural decline, then it endangers everything Marshall Islanders value most. Yet by framing it as cultural decline, belief in that traditionalist discourse is bolstered and social solidarity is thereby strengthened. Climate change makes Marshall Islanders less secure, but the *idea* of climate change makes them more so; just as pervasive decline itself makes Marshallese "tradition" less secure while the *discourse* of pervasive decline makes it more secure by affirming its value and demanding its protection. The situation has parallels in the West. The left fears climate change, for it endangers the already endangered, but it also adores climate change, for it exposes the failures of unfettered accumulation and unquestioned progress. The right fears terrorism, for it imperils the cherished homeland, but it also adores terrorism, for it proves the necessity of military resolve and the moral bankruptcy of the West's opponents.

Given the ideological benefits that a society may accrue from physical dangers, climate change can be taken as an opportunity rather than a burden. To say that Marshall Islanders are facing climate change is a double entendre. The inhabitants of low-lying islands are often presented as passive with regard to climate change, merely *facing* future destruction (Farbotko, 2005; for an example, see Smallacombe, 2008: 73). This book has shown how they can also *face* climate change, face up to it, respond to it. This case study thus carries fairly optimistic implications with regard to humankind's ability to solve, cope with, or harness climate change. Humanity's sluggish response to climate change may not be an inevitable result of the various liabilities summarized in Chapter 3, nor is it due simply to inertia or an unwillingness to face danger. It may rather stem from the fact that concern entrepreneurs have not yet discovered ways to make climate change appeal to various ideologically diverse communities. While some may subscribe to worldviews so unambiguously contrary to the notion of climate change that no amount of rhetoric will convince them, there are surely other, as yet unconvinced groups that could be brought on board if only the appropriate arguments and imagery were used.

Both activists and scholars are fond of portraying anthropogenic global warming as a momentous idea that will, and must, radically alter our worldview (Beck, 1992; Chakrabarty, 2009; Donner, 2007; Hulme, 2010c; Jasanoff, 2010; McKibben, 2006[1989]). Life in the Anthropocene erodes the cornerstone of western modernity, the separation of nature from culture. It collapses the gulf between future and present. It makes global citizens, or cosmopolitans, of us all. It proves the folly of endless growth, of unrestrained capitalism, and some would say of modernity in general. Human colonization of the atmosphere may go so far as to upset the divine and unseat God (Donner, 2007; McKibben, 2006[1989]). Perhaps these radical shifts of perspective are sorely needed. Perhaps climate change ought to be an impetus to rethink and remake ourselves. But for the activist who has taken on the more modest goal of raising awareness and concern, the idea that climate change demands a philosophical revolution can only be seen as an invitation to despair. It is enormously difficult to dislodge people's deepest convictions; if that is what successful climate change communication requires, it is most likely doomed from the start. Happily for the science communicator (but unhappily for the more radical cultural agitator), the idea of climate change is much easier to swallow than thinkers like Bill McKibben, Simon Donner, and Sheila Jasanoff assume. For better or for worse, Marshall Islanders have taken on board the idea of climate change without threatening any of their core convictions. If this can be done in one of the world's most threatened countries, perhaps it can be done anywhere; and the literature on confirmation bias and belief resilience, on the cultural appropriation of risk, on the molding of climate change discourse to fit prior commitments, hints that this is indeed the case.

Certain fears about the intractability of climate change are thus partially allayed. Orlove, Wiegandt, and Luckman (2008: 14) worry that climate change impacts, being caused primarily by distant agents, will inspire a sense of victimization and therefore disempowered passivity. The Marshallese case suggests that such problems may be overcome by casting the threat in locally resonant terms, the result of which may be the very opposite of victimized apathy—and may even err on the side of assuming too *much* responsibility. Sheila Jasanoff (2010) similarly worries that climate change as a universal, value-free scientific discourse is incompatible with the situated, normative interpretations that might inspire real concern. She worries, too, that the challenges that climate change poses to prior human understanding of nature and society create an "[intellectual] climate that renders obsolete important prior categories of . . . experience" (Jasanoff, 2010: 233) and "cuts against the grain of ordinary human experience, the basis for our social arrangements and ethical instincts" (Jasanoff, 2010: 237) in ways so numerous and deep that they may take centuries to accommodate. These fears are reasonable but perhaps not as discouraging as they appear. No matter if the scientific idea of climate change is presented in universalistic ways shorn of local meaning; communities will *make* it locally meaningful. Global action on climate change may therefore be easier than we have feared—not because of a global consensus on the meaning of climate change but rather owing to a plurality of local imaginations of climate change, each one spurring the same needed responses (mitigation, adaptation) but for differing reasons: a necessary commonality of action without an unnecessary commonality of thought—a humanity, to invert the famous phrase, thinking locally and acting globally.

A TRAJECTORY THEORY OF RISK PERCEPTION

Exactly as Mary Douglas would predict, Marshall Islanders have transformed climate change from an abhorrent *danger* to an appealing *risk*. The general cultural theory of risk has been corroborated. Its easy applicability to a threat (climate change) and a society (the Marshall Islands) for which it was not specifically designed is compelling evidence in the theory's favor. But Douglas's theory has fallen short in another regard. It has failed to draw attention to what is, I have argued, the most important prior commitment, namely the discourse of seductive modernity and cultural decline. A few authors have hinted at the influence of such *trajectory narratives* on risk perception, but the analysis has never been pushed very far. Mary Douglas briefly discusses optimistic and pessimistic tendencies in societies but does not elaborate (Douglas & Wildavsky, 1982: 121–122, 190–191). Richard Norgaard suggests that economists discount environmental threats because the Enlightenment faith in material progress is at odds with the degradationalist counternarrative of environmentalism (Norgaard, 2002); Aaron McCright briefly makes a similar point regarding the right-wing reception of climate change (McCright, 2011). Jonathan Jackson suggests that alarming crime statistics are more credible and resonant to publics who believe that society is on the decline; crime is used as a moral barometer, and the publics' statements that they fear criminal harms are intended to express a belief in the downward trajectory of social morals (Jackson, 2006). Marian Tulloch similarly finds that elderly Australians' fear of crime is mediated by their belief in a deteriorating Australian nation (Tulloch, 2000). None of these authors, however, extends these insights outside of the realm of their case study or proposes any sort of general theory linking trajectory and risk. By and large the connection has been overlooked. I therefore propose a "trajectory theory of risk perception," a more general statement of the trajectory theory of climate change attitudes that I have proposed elsewhere (Rudiak-Gould, 2012b).

A trajectory theory of risk perception would assert that societies, factions, or individuals (for the sake of brevity I will simply say "societies") with a vigorous discourse of decline will select more risks for concern, because *any* present or future risk confirms the preexisting belief in a worsening universe. Such societies will be especially likely to select *apocalyptic* threats for societal attention and any risk that implies a benign, dangerless former state of the world. Environmentalist threats will thus be highly credible to such societies, for the reasons outlined in Chapter 3. Conversely, societies with a vigorous discourse of progress will be biased against believing in *any* present or future risk, especially apocalyptic risks and those that presume a former, better way of things. Such societies will select fewer risks for attention and be less credulous of danger discourses in general. They will, however, be credulous of *past* risks and willing to select *mild* and *surmountable* present and future risks for societal attention—risks over which the steady march of progress will prevail, proving its durability, rather than ones on which it founders, proving its fragility. Such societies may also being willing to credit risks that posit sudden, intermittent bursts of danger rather than a gradual decline in safety. *Responses* to threats will proceed accordingly: regress societies will favor radical responses (the world being in a downward spiral, drastic action is required) and will be more eager to take up threats that seem to require such fundamental reform; meanwhile, progress societies will favor shallower responses (the world being upward bound, a radical change of course is unnecessary or dangerous) and will be reluctant to take up threats that seem to demand radical change.

Regarding global climate change more specifically, a trajectory theory of risk perception would predict that regress societies will eagerly embrace climate change, will interpret it in apocalyptic terms, and will favor radical responses, while progress societies will disbelieve climate change or regard it as only mildly dangerous, calling for a technological or bureaucratic quick fix. This prediction seems to be borne out by left-wing versus right-wing responses to climate change in the West. This apparent corroboration could be more rigorously demonstrated by conducting studies similar in methodology to Feygina and colleagues (2010) and Feinberg and Willer (2011); the trajectory theory of risk perception would predict that a large part of the variation between individuals in their attitudes to climate change—belief in it, preoccupation with it, assessment of its severity, and the drastic nature of proposed responses to it—will be accounted for by the individual's preexisting belief in regress or progress. Such variables, my theory suggests, might account for much of the variation previously accounted for by variables such as a just-world belief (Feinberg & Willer, 2011), system justification (Feygina et al., 2010), and liberalism/conservatism (McCright, 2011).

In Pacific societies, the trajectory theory of risk would predict that societies with a strong narrative of traditionalist decline would select and interpret risks in the aforementioned manner of decline societies: they will be highly receptive to climate change and other degradationalist risks and frame them in traditionalist terms. Meanwhile, Pacific societies with a vigorous sense of religious progress or modernist progress will reject or downplay such risks. Within single Pacific societies, individuals particularly committed to religious discourse (such as church authorities), to modernist discourse (such as government officials), or to traditionalist discourse (such as chiefs) will subscribe primarily to the corresponding moral trajectory narrative and will respond to threats appropriately. Thus the theory predicts that different individuals and factions in a Pacific society will find themselves at odds over climate change for the many of the same reasons they are at odds over other issues. It might predict, for instance, that the traditionalist chiefly establishment in Fiji will embrace the idea of climate change, while Methodist church authorities, more attached to an antitraditionalist ideology or a belief in religious progress, will regard climate change with more skepticism.

The influence of *blame* narratives, as an aspect of trajectory narratives, could be added to the trajectory theory of risk perception. In progressivist narratives there is no need for a strong notion of blame: the world is getting better, so no one need be accused. In decline narratives, however, blame figures prominently. Crudely, we could distinguish between in-group-blaming regress narratives and out-group-blaming regress narratives. An extended trajectory theory of risk perception would predict that the account of blame in the narrative of moral trajectory will influence what risks are selected for attention and shape the moral interpretation of those risks. In-group-blaming regress societies (like the Marshall Islands) will select risks that accuse the in-group, reject risks that do not, and interpret risks that are ambiguous vis-à-vis blame (such as climate change) as being primarily perpetrated by the in-group. Out-group-blaming regress societies will select risks that incriminate disliked foreign groups, reject risks that accuse the in-group, and interpret ambiguous risks like climate change to implicate an out-group, in particular one against which the group already harbors grievances. The extended trajectory theory of risk would also suggest that certain documented cases of climate change self-blame, such as among rural sub-Saharan Africans (BBC World Service Trust, 2010; Patt & Schröter, 2007), might stem in part from prior attachment to an in-group-blaming decline narrative.

To begin to cross-culturally test and apply the theory I have outlined here is beyond the scope of this book, but I will briefly examine a particular case study that appears to corroborate it. William D. Smith (2007) provides a portrait of climate change attitudes among indigenous Totonac farmers in Mexico. A key informant opined that since the advent of moneyed life and cash-cropping, the social and physical worlds have declined; indeed, for this informant the "two" worlds were one. A deteriorating, destabilizing climate, which he had himself observed, was part of this more general decline. While fully aware of the superior power and wealth of the neighboring United States, the Totonac man placed primary responsibility for this decline on local people's desertion of customary faculties of "presence of mind" and proper human-environment interactions. As a result, the informant advocated a "back to basics" (W. D. Smith, 2007: 222) approach to tackling both climatic changes and other local woes. In all ways this appears to be another case of the subsumption of climate change into a preexisting discourse of self-inflicted decline. Smith does not theorize this aspect of his case study in depth or propose any sort of trajectory theory of risk perception, nor does he provide more direct evidence that belief in decline contributes to concern about climate change. Even so, his case study seems remarkably similar to the Marshallese case, hinting that a trajectory theory of risk perception might have currency and insight even in a radically different cultural context.

These predictions could be tested both quantitatively, with the methodological tool kit of psychology (see Feinberg & Willer, 2011; Feygina et al., 2010) and sociology (see McCright, 2011), or qualitatively, with the ethnographic approach favored by most anthropologists. Many insights might result from such a research project. Mark Nuttall (2008: 51) states that the usual indigenous perception is to regard climate change as a severe threat perpetrated by foreign groups; he documents an exception in Greenland, where indigenous people often see profoundly positive opportunities in warming trends. I have documented another exception to Nuttall's generalization: an indigenous group that tends to regard climate change as severe but blames the in-group for it. Testing and applying the trajectory theory of risk perception might reveal that Nuttall's and my "exceptional" cases are not in fact unusual; they may even be common. If indeed there are many indigenous groups, factions, or individuals who subscribe to narratives of religious or modernist progress, then it may not be unusual, after all, to find indigenous people who doubt the reality of climate change or manage to reframe it as an opportunity; and if there are indeed many indigenous groups who subscribe to in-group-blaming decline narratives, then self-blame for climate change may be quite common as well. Cross-cultural testing of the trajectory theory of risk perception could uncover both the existence of these patterns and the reasons for their existence. In these and other ways, research inspired by the trajectory theory could enrich our understanding of risk perception, the role of moral trajectory narratives in society, and human responses to the threat of climate change.

LOOKING FORWARD

I end this book with a look to tomorrow, a series of predictions for the future of climate change perceptions in the Marshall Islands. As impacts intensify, climate change will become the new radiation, the catch-all explanation for negative change. Diabetes will come to be blamed on climate change. The ranks of hyperbelievers, those who see signs of climate change in nearly everything, will increase, and their trust of scientists will swell; *scientists as soothsayers* will become the dominant view. As two vertices of the triangulation—firsthand evidence, scientific trustworthiness—are satisfied, the third vertex—biblical exegesis—will be needed more than ever by any remaining disbelievers. Those skeptics, if they still exist, will use the Genesis argument for all it is worth and may even extend it with a passage as yet undiscovered by Marshall Islanders in the context of climate change, a passage in which God promises not only that the sea will not rise but that nature in general will abide:

> The Lord . . . said in his heart: "Never again will I curse the ground because of man. . . . And never again will I destroy all living creatures, as I have done. As long as the earth endures, seedtime and harvest, cold and heat, summer and winter, day and night will never cease" (Genesis 8:21–22).

For a climate change skeptic, the generality of this prophecy neatly undermines the promiscuous corroboration of climate change, the convenient broadness of the word *mejatoto*. Believers, in riposte, will need to step up their own scriptural exegesis, comparing disbelievers to those who failed to heed Noah's warning, and perhaps extending that argument by finding yet another passage, undiscovered until now, one that handily references sea level rise and the inundation of islands: "Every island fled away and the mountains could not be found" (Revelation 16:20).

In-group blame will endure, for it lies on ideologically solid ground. But Marshall Islanders are not entirely averse to out-group blame, so their universal blame for climate change could start to emphasize other culprits as well. In particular, if locals, already frosty toward Chinese immigrants, learn that China is now the world's largest carbon polluter, blaming that country for climate change could become an attractive option.

As climate change turns a known environment into an unknown one (Christine Pam and Rosita Henry, personal communication), as the ecology and the weather become more treacherous and unpredictable, all of the Marshallese world will turn from Malinowski's lagoon to Malinowski's ocean: supernatural protection will become a psychological necessity. The stimulus will grow to revitalize magical methods of thwarting natural disasters, such as the powers of the Ripako. But the very existence of the dangers that this magic aims to nullify will be seen as evidence of its ineffectuality, of the decline of magical potency. Supernatural means to stay safe will become ever more alluring, ever more unachievable.

If and when wholesale resettlement occurs, already passionate cries of cultural breakdown will acquire new heights of fervor. The conceptual shift, if not the emotional one, will be easy: now-salient differences between the outer islands and the urban centers—the former the site of tradition retained, the latter the site of modern decay—will be elided, and the Marshall Islands as a whole will become synonymous with tradition, the good life, the past. As the last island sinks and the last family leaves, Marshall Islanders will realize that they have been forced to sell their country for a way of life that they repudiate.

Notes

NOTES TO THE INTRODUCTION

1. By calling these questions "debates," I do not mean to imply that there is serious scientific uncertainty about the reality of climate change or its largely anthropogenic origins. The former is now unequivocal, the latter extremely probable. I do mean to point attention to the fact that public debate still persists on these issues in many societies (Brugger, 2010; McCright, 2011; Sarewitz, 2004; Shearer, 2011). In the case of whether climate change is occurring and whether it is human-made, perhaps eventually the debate will cease, for here we are dealing with empirical questions that can potentially be settled beyond a reasonable doubt by evidence. In the case of who is to blame for climate change (which is not the same as the question of who is causing it), the debate will likely never be resolved because different normative commitments are at stake. In the case of what is to be done about climate change, this too will not be resolved by more education because normative debates for which scientific evidence offers no remedy are involved and even the empirical issues are extraordinarily complex and beyond our powers of firm prediction. All three debates are alive in the public sphere, no matter how much scientists have attempted to quell them with ever greater scientific knowledge and public education (Sarewitz, 2004).
2. In this book, the term *mitigation* is used in its more limited sense—the reduction of the *causes* of climate change (greenhouse gas emissions)—rather than in its extended sense, encompassing the reduction of the negative *impacts* of climate change.
3. In this book, the term *adaptation* is meant to include any strategies to cope with climate change *other than* mass depopulation of the archipelago. There is ambiguity in the term: migration may be considered either a form of adaptation or a failure thereof (see Farbotko & Lazrus, 2012; Orlove, 2005). In this book, corresponding to widespread Marshallese views on the matter (see Chapter 5), nationwide evacuation is considered to be in opposition to adaptation rather than a part of it (see Barnett & Webber, 2010).
4. My survey respondents were adults in the D-U-D (downtown) area of Majuro. In order to reach the largest audience, I conducted the survey in the Marshallese language (except for a very small number of individuals who preferred to speak in English) and did so orally so that English competence and literacy were not requirements for participation. I employed convenience sampling: I walked the streets in every neighborhood and administered the survey to people willing to participate. All of the respondents were individuals I had never met and who did not know who I was,

thus ensuring that people were not biased by knowing that I was studying climate change. While the lack of a random sample may skew my results, the distribution of age, gender, education, and religious domination are close to national statistics, and nothing in my fieldwork led me to believe that this survey misrepresents the Marshallese population as a whole. To the first 100 respondents I administered a version of the survey in which observations of environmental change were elicited before asking about the respondent's familiarity with the scientific concept of climate change; this was to avoid priming the subject to report environmental change. The final forty-six surveys were administered in the opposite order: the subjects were asked first about the scientific concept of climate change and then about their observations of local environmental change.

5. In both surveys, not all participants responded to all of the questions, so I report the number of respondents for a particular question whenever I report percentages of particular answers to that question. Throughout the book, percentages do not always add up to 100 because I round to the nearest unit.
6. Although I employed a mixed-methods approach, it must be said that this book relies in a large part on elicited, verbal data, whether those data come from structured surveys or from more open-ended interviews. This was a matter of practical necessity. Although I was able to record various unsolicited statements and spontaneous practices relating to climate change, these data by themselves would have been insufficient for this research project; interviews and surveys were necessary. While the pitfalls of elicited verbal data are numerous and well known (Berreman, 1962; Gombrich, 1971: 37–38; Lanman, 2007; Ray, 1990; Southwold, 1983: 135–136; Whitehouse, 2007), it remains indispensable to the ethnographer, especially when the topic in question is a chiefly future-oriented threat. I employ a number of methods to minimize the danger: I "triangulate" (Jack & Roepstorff, 2003: vii), whenever possible, between different sorts of data: comparing prompted verbal statements with unprompted ones, and both kinds of statements with behaviors. I analyze the unstated assumptions and conceptual schemes underlying informants' statements rather than taking a face-value, just-ask-and-they-will-tell approach to interview data. I minimized the biasing effect of impression management (Goffman, 1959; Leary & Kowalski, 1990; Leary, 1995) by feigning a lack of opinion on the issues at hand until the informant had told me his or her opinions, by avoiding announcing outright that I was studying climate change (following Mortreux & Barnett, 2009: 107), by relying on open-ended rather than forced-choice questions (following Ray, 1990), and by usually interviewing people privately and one on one, assuring them (with the exception of certain public figures) that their statements would be recounted only anonymously.

NOTES TO CHAPTER 1

1. For more on chiefs in discourse and practice, see Carucci (1997a) and Walsh (2003).
2. Blackbirders were sailors who kidnapped Pacific Islanders to be used as plantation laborers in Fiji, Samoa, and elsewhere. Few returned alive.
3. Albeit in the case of some remote atolls, not until well after World War II (see Carucci, 2003: 62).
4. For details of Marshallese land tenure systems, see Kiste (1974) and Mason (1987).

5. Marshall Islands Journal, June 19, 2009, p. 33.
6. There are striking parallels between this legend and the previous one. In both stories, Ļetao introduces a powerful force that can both help and harm. In both stories Ļetao's trick hinges on the dual nature of fire, bringer of both cooked food (life) and destruction (death). Thus Ļetao is, like Maui, both creator and destroyer.
7. No Marshall Islander has explicitly told me this interpretation of the legend. I have devised it myself. However, it fits so well with the narrative of cultural decline and the conscious connection made between Ļetao and America that I suspect many Marshall Islanders have it in mind in telling or hearing the legend. For slightly different versions of this legend and other analyses of the cultural significance of Ļetao, see Carucci (1997b: 148–149), Kramer and Nevermann (1938: 239–250), and McArthur (2000). For other Ļetao legends, see Kelin (2003).
8. When I asked survey respondents (with sixty-five responses to this question), 90% specifically blamed Marshall Islanders, while only 5% blamed outsiders, such as Americans. (Only 5% blamed people in general.)
9. I use this analogy merely as a device to render the Marshallese narrative more intelligible and familiar to readers. But the parallels between the demise of tradition and the expulsion from Eden are so striking that one cannot help wondering if this is not accidental. Perhaps Marshallese traditionalist narratives were Biblically inspired (as Tomlinson [2009] suggests for Fiji); some locals do include Christian elements, such as the prediction of coming End Times, in their narratives of seductive modernity. But there is no direct evidence of a link. I never heard any Marshall Islander explicitly liken cultural decline to the Fall or America to the Serpent.
10. Yet another historical narrative sometimes encountered in the Marshall Islands points to progress in the form of economic development. John Silk, the Minister of Resources and Development, told me in 2007, "I think the lives of people have improved. And one of the indicators I look at is . . . what the government is trying to do to uplift the economy, and try to improve the lives of people. I look at the water catchments that they're putting up. I look at the renovations of the schools and dispensaries. I look at the program that the government is now working on to bring solar power to the outer islands. . . . I look at the field trip ships that the government provides. . . . So if you look at the overall picture, you'll see that we've made some progress." This narrative is popular in the halls of government (though government spokespeople also espouse traditionalist decline at times). See Walsh (2003) for an in-depth treatment of this particular local view of Marshallese history.
11. Other scholars of the Marshall Islands have written extensively on Marshallese views of Americans (Carucci, 1997b, 1989; Kiste, 1974; McArthur, 2000; Walsh, 2003) and on the figure of Ļetao (Carucci, 1997b: 148–149; Kramer & Nevermann, 1938: 239–250; McArthur, 2000), but they have touched on Marshallese metaculture (see Carucci, 1990, 2003; Kirsch, 2001; Poyer, 1997: 71–72) and narratives of decline (see Carucci, 1997b: 167; 2003: 69; Kiste, 1974: 148; Poyer, 1997: 72–73, 81) only in passing. The previous literature has also neglected to fully integrate these three topics with each other: Marshallese views of themselves, of outsiders, and of cultural change. For Marshall Islands scholars and Oceanists, this is the ethnographic contribution of this book.
12. As Carucci observes of the Marshall Islanders of Eniwetok and Ujelang atolls, "At the same time that overnight Americanization was abhorred, somehow the community could not resist its charms" (Carucci, 2003: 69).

NOTES TO CHAPTER 2

1. Ninety-seven individuals answered this question, with 130 total mentions of information sources.
2. Sixty-one high school-aged respondents to the Marshall Islands High School survey in 2009 reported having heard about climate change through the following sources: school (39%), television (21%), radio (20%), word of mouth (18%), and newspaper (2%).
3. Global English-language media witnessed a similar rise in climate change coverage during the first decade of the twenty-first century, but it appears to have begun earlier, in 2004 rather than 2007 (Boykoff & Roberts, 2007). For more on climate change reporting in Pacific Island nations, see Jackson (2010).
4. Majuro has also been the site of several climate change-related international conferences, including the Micronesian Traditional Leaders Conference in October 2008; the Regional Climate Change Adaptation Workshop in April 2009; the Micronesian Chief Executives' Summit and Micronesian Presidents' Summit in July 2009 (with theme "Climate Change and the Energy Challenge: Proactive Leadership for a Resilient Micronesia"); and the Pacific Climate Change Roundtable in October 2009. These meetings were conducted in English and thus disproportionately reached the educated elite, but they also trickled down to some extent to other Marshall Islanders through radio coverage and word of mouth.
5. Simply asking informants if they have heard that scientists claim the ocean is rising and the weather is changing due to human activities risks the acquiescent response bias. Asking if the person has heard of the Marshallese translation *oktak in mejatoto* risks an overestimation of climate science awareness because this Marshallese phrase sounds so familiar and untechnical that respondents would say they had heard of it even if they had never heard about global climate change. Asking instead if the individual had heard English phrases like "climate change," "global warming," and "the greenhouse effect" excludes those who are well aware of the scientific discourse of anthropogenic climate change but have not heard or do not remember the English phrase.
6. For comparison, the 2006 Nielsen Global Omnibus survey gives figures for having "heard or read anything about the issue of global warming" ranging from 99% for the Czech Republic to 75% for Malaysia, with the US and the UK coming in at 83% and 92%, respectively (Boykoff & Roberts, 2007).
7. Opponents employed the same rhetoric to support the opposite agenda: they argued that women ought not to have such a right because they were not traditionally granted it.
8. The same invocation of Genesis to deny the reality of climate change has been documented in Kiribati (Teuatabo, 2002: 89) and Tuvalu (Lynas, 2004: 117; Mortreux & Barnett, 2009; Paton & Fairbairn-Dunlop, 2010), low-lying Pacific nations facing much the same future as the Marshall Islands.
9. It is not surprising that some Marshall Islanders would find a way to dismiss the possibility of inundation. Although, as I argue in this book, locals have largely accepted the idea of climate change, the prospect of nationwide destruction is nonetheless a horrific thought, and there is some obvious comfort in denying it. The Genesis argument is the most common justification for disavowal, but others exist as well. A middle-aged woman in Majuro told me, "I don't believe [climate change will destroy the country], because if there is so much sunshine [from global warming], it will take the water away." A scientifically educated man speculated that as the temperature of the water increases, the evaporation will increase as well, canceling out sea

level rise, or the evaporation will create a mist that cools the earth. A few islanders said that the country would not be destroyed by climate change because the ancestors had cast spells on the islands to protect them from wave damage (see Chapter 3). Numerous studies have demonstrated people's ability to discount the possibility of being harmed, rationalizing this comforting denial through whatever discourses are available, especially if they feel that the threat is outside of their control (Edelstein et al. 1989; Grothmann & Patt, 2005: 203; Grothmann & Reusswig, 2006: 106; Kroemker & Mosler, 2002; Lehman & Taylor, 1987). Similar climate change disavowal strategies have been documented in other cultures. Carla Roncoli reports that some farmers in Burkina Faso predicted ample rainfall after a drought on the grounds that God would not test people past their breaking point (Roncoli et al., 2002). Faith in God's protection against climate change has been reported in low-lying island nations other than the Marshall Islands—indeed, in all other sovereign nations at risk of total uninhabitability due to climate change: Tuvalu (Mortreux & Barnett, 2009: 109–110), Kiribati (Teuatabo, 2002: 89), and the Maldives (Toomey, 2009). US Congressman John Shimkus publicly stated in 2009 that global warming could not be real because of God's promise to Noah that "As long as the earth endures, seed time and harvest, cold and heat, summer and winter, day and night, will never cease" (Genesis 8:22) (Mail Foreign Service, 2010).

10. This does not follow the Bible to the letter, because in the Genesis passage God promises not only that he will not flood the earth but that a flood will not happen, full stop. We might expect some locals to look more closely at the passage and reject the refutation on these grounds, but no one I met did so.
11. The same argument has been put forth by the Pacific Council of Churches in their Otin'taii Declaration on climate change (Pacific Council of Churches, 2004).
12. Marshall Islanders may not be the only indigenous people to flirt with this particular biblical interpretation of climate change: one woman in the Porgera Valley of the Papua New Guinea highlands said that a recent warming trend was "the fire that Jesus was bringing from heaven to destroy the world" (Jacka, 2009: 205).
13. Some Christians in the West hold the same interpretation of climate change (Hulme, 2009: 154–5; also see Jennaway, 2008).
14. There are occasional exceptions. An American missionary in Laura has told locals that he is not sure whether climate change is true because not all scientists agree about it. Yumi Crisostomo, during her presentation at the WUTMI Executive Board Meeting in August 2009, stated that 95% (not 100%) of scientists believe that climate change is occurring. But Marshall Islanders, unlike some climate skeptics in the West, have made nothing of the idea that anthropogenic warming is "just a theory" that not all experts accept.
15. Throughout this section, I could attempt to separate two distinct meanings of the term *trustworthiness*: trustworthiness as expertise (knowing the truth) versus trustworthiness as honesty (telling the truth). For a few reasons, I do not attempt this clean separation. The two kinds of trustworthiness are closely intertwined in Marshallese thinking. As we will see, expertise becomes conflated with honesty in the view of *scientists as God's chosen ones*, and expertise becomes conflated with dishonesty in *scientists as God's rivals*. Moreover, the important question for Marshall Islands is whether scientific statements on climate change are true, not whether their untruth stems from ignorance or duplicity.
16. When I asked locals why global climate change was occurring, people often said, "You would know!" as if being American entailed, at the very least,

some knowledge of science. In general, though, locals did not assume that I was a scientist simply because I was American. Despite being a foreign seeker of knowledge, in a crucial regard I violated the definition of a scientist, because my techniques and aims were intuitive: I spent time with people and talked to them in order to learn how they live and think. People found this perfectly comprehensible and relatable. How different this ethnographic endeavor was from the activity of a physical scientist, with highly specialized methods and mysterious devices requiring years of training to comprehend. So *jaintiij* does not simply mean a foreigner in pursuit of knowledge so much as a foreigner with powerful and mysterious tools with which to examine and influence the world.

17. This relates to a puzzle in Marshallese ideology: is America a Christian nation, or an un-Christian one? Marshall Islanders remember Americans as the bringers of Christianity; but now many American visitors to the country are not Christian and in fact break Christian rules by working on the Sabbath. In a neat microcosm of this paradox, American missionaries originally demanded that local women dress conservatively in muumuus, covering their breasts, ankles, and shoulders (in fact, the most common word for American, *ripālle*, literally means "those who cover" or "those who wear clothes"); but now it is American visitors who are scantily clad, while Marshallese women guard their modesty. The roles have reversed: Americans were once the saviors of Marshallese sinners, and now Americans sin while Marshallese people are saved. The paradox was neatly encapsulated on Gospel Day on Ujae in 2003, when locals celebrated and reenacted the advent of missionaries in the Marshall Islands. I was chosen to play the captain of the missionary ship in this reenactment. As the only American on the island, I was the natural choice for the role. But I was also the island's only non-Christian. I was thus, simultaneously, the least and most missionary-like individual on the island. Locals often assumed that I was Christian and would ask me where, in America, I attended church. When I said, "Nowhere," they were surprised, and one woman admonished me, "But it was you Americans who brought religion to us Marshallese people!" Locals sometimes say that scientists (and Americans) are Christians while other times declaring they are not. The paradox, we can see, is not easily resolved.

NOTES TO CHAPTER 3

1. By "believing in climate change" I mean accepting (as shown through private statements, public statements, and behavior) that the usual version of climate science presented to Marshall Islanders is basically right—that particular changes (sea level rise, warming, increased drought) are afoot across the planet, caused by the worldwide human use of technology, and expected to have severe consequences to the Marshall Islands in the future. The concept of *belief*, of course, has been critiqued by anthropologists as a western, Christian, or academic preoccupation not shared by locals (see Graveling, 2010; Ruel, 1982; Southwold, 1983: 131–135), and some environmental anthropologists object to its use in the realm of climate change perceptions. But with regard to Marshallese views of climate change, belief is not my preoccupation but theirs. As we saw in the previous chapter, when locals speak about the threat, they often express belief or disbelief. In fact, the Marshallese language has a word for "believe" or "belief," *tōmak*, used much as in English and applied to a variety of topics, including climate change, religion, scientific expertise, cultural preservation, and so forth; one can *tōmak ilo*

("believe in") God as well as *tōmak* ("believe") that tomorrow will be rainy or *tōmak ilo* ("trust in") scientists' proclamations. *Belief* is, like *culture*, a "deeply compromised word that I still can't do without." (Clifford, 1988:10) (See Lanman [2008] for a defense of the academic concept of *belief*.)

2. This may stem from an older practice of marking particular trees or garden plots as forbidden to others (see Kotzebue, 1821: 58, 72, 102).
3. Also spelled Dri-Bako, Di Pako, Ribako, and Ribōgo.
4. Strictly speaking, the Ripako are not extinct. Ripako clanspeople live on Arno (Petrosian-Husa, 2004) and Eniwetok (Carucci, 1993: 165; 2003: 58) atolls and both groups claim that the clan originated on their home atolls. Descendants of the Jaluit Atoll Shark People, however, consider the clan to have originated on Jaluit Atoll, say that the clan is now extinct, and are unaware of the existence of Ripako clanspeople on other atolls. It is this second account, the story of the Ripako as told by those with ties to Jaluit, on which I focus.

NOTES TO CHAPTER 4

1. Tibetan villagers in Yunnan province were quite sure that human actions were at the root of recently observed local climate change, even though they were uncertain about how exactly this was caused (Byg & Salick, 2009: 165). Blaming environmental change on humans is extremely intuitive, to the extent that no clear mechanism of causation need be described.
2. The percentages add up to more than 100% because a few respondents gave more than one answer.
3. It would not, however, be easy for Marshall Islanders to blame "the West," because they have no category for this. The closest equivalent in the Marshallese language is *ripālle*, which can refer to all Caucasians or people of European descent, but whose much more common and precise meaning is "American." Meanwhile, the term *laḷ ko rōḷḷap* refers to any large country, including non-Western ones such as China and Japan. Westernness per se has never been an important concept for the Marshallese, since their colonial masters have been East Asian as well as Euro-American. The Marshall Islands has been colonized by *laḷ ko rōḷḷap*, not by the West.
4. One could also argue that people adopt in-group blame because local sources of pollution are more noticeable than foreign ones (see BBC World Service Trust, 2010). This requires a folk theory of climate change causation in which causes and consequences are locally specific. A small number of Marshall Islanders do hold such theories. For instance, there is a folk theory that the atmospheric "blanket" (*kọọj*) can be ripped in some places more than others, so pollution issuing from the Marshall Islands damages only that country's section of the atmospheric blanket. But such theories are rare in the Marshall Islands, meaning that this line of reasoning cannot explain much.
5. Another locally unpopular group, of course, is the Chinese, but it has apparently not occurred to Marshall Islanders to blame them for climate change. If the fact is disseminated that China is now the world's number one greenhouse gas emitter, perhaps locals will begin to blame Chinese people. Unlike in the case of Americans, there are no positive views of the Chinese to temper the negative ones, no sense of loyalty counterbalancing the sense of resentment. They could therefore be tempting targets of accusation.
6. The government (in this case the EPA) sponsored this event, but I include it in this list of grassroots mitigation efforts because it involved a large amount of participation and support by nongovernmental Marshall Islanders. The

same can be said of the other events in this section that were sponsored or cosponsored by Marshallese government bodies.

NOTES TO CHAPTER 5

1. In all of my time in the country I met only one individual who told me that climate change, in and of itself, had convinced him to leave the country. In 2007, a government worker, who spoke on condition of anonymity, said that he was entirely sure that in fifty years the country would be fifty feet under water. In response to my statement that I hoped Marshall Islanders might find a way to stay in the country, he said, "Don't hope. You can't deny God's word. It's in the Bible." He likened the event to the biblical flood and vowed that he would leave the country as soon as he received his government pension. I met two other individuals who said that climate change fears had constituted an important part of their decision to move. A young Majuro man from a chiefly family was convinced in 2012 that the country would be destroyed by climate change and partly for that reason expressed a wish to move to Hawaii. A politically active man in his thirties said in 2012 that he was considering leaving the country in ten years in part because of climate change but also because of the poor outlook for the political leadership.
2. This is true even in Tuvalu (Mortreux & Barnett, 2009; Paton & Fairbairn-Dunlop, 2010), a country that has become synonymous in the international media with climate change vulnerability and that is often erroneously reported to be in the process of evacuation or already deserted—for instance, in the widely viewed *An Inconvenient Truth*.
3. This migration provision could disappear in 2023 when the Compact of Free Association is renegotiated.
4. The forced migration literature echoes this sentiment. Unwanted resettlement often ravages a community's physical, psychological, and social well-being for decades (Campbell et al. 2005; Eisenbruch, 1990; Harrell-Bond, 1999; Lee, 1990; Moore & Smith, 1995: 115; Mortland, 1994; Neumann, 1997; Oliver-Smith, 2009: 126; Silverman, 1971). The suggestion that mass resettlement is an unexceptionable solution to climate change has been debunked by ethnographically sensitive scholars (Adger & Barnett, 2005; Crate, 2008: 573; Mortreux & Barnett, 2009; also see Kempf, 2009: 200–201; Kirsch, 2001).
5. When I told MICS Executive Director Albon Ishoda this interpretation in 2012, he immediately understood and agreed, saying, "Look at them [the Bikinians] now. They live in Kili. . . . They don't know how to fish. . . . They don't grow food. They're all dependent on the money and the food from the USDA. . . . And they've lost a big part of who they are. . . . Bikini people were very known for their navigation skills. . . . Not any more. . . . [Now] they navigate by the horsepower. [Laughs.]. . . . We've been exiled once before, exiled within our own country. Now they're thinking about exile . . . to another country. . . . and I think that's part of the reason why at the highest levels of our government there's a no-move policy."
6. Enemanit-Latuma Extended Family Association.
7. There is some disagreement as to the Linnaean equivalents to the Marshallese terms. For consistency I use those species identifications found in the Marshallese-English Dictionary (Abo et al., 1976).
8. The government has expressed its commitment to adaptation in climate change policy documents. The Climate Change Roadmap 2010—drafted in August 2010 by Philip Muller (the Marshall Islands ambassador to the

United Nations), Japan ambassador Jiba Kabua, and representatives from government ministries and NGOs—advocates revegetating coastlines, building seawalls, safeguarding water resources, improving disaster management, and protecting coral reefs, among other measures, and aims to fast-track the country's access to the many sources of adaptation funding now available; it was approved by the cabinet the following month. The government now presents its participation in the Micronesia Challenge initiative (a multinational commitment to protect 30% of near-shore marine areas and 20% of land areas) as partly or primarily a climate change adaptation measure, not merely a conservation project (Journal, 2010d), and the EPA now attempts to limit coastal dredging partly on the grounds of reducing climate change vulnerability (Journal, 2011f). UN Ambassador Philip Muller, with the backing of the president's office, announced in late 2010 an intention to seek $20 million in foreign donations to begin engineering work on a 5-kilometer-long seawall to protect much of Majuro's urban center (planned to run from Delap district to the Marshall Islands High School, in Rita district), along with filling small bays with landfill. The government shortly thereafter requested funding from the United Nations for reverse osmosis water purification technology as a climate change resilience measure (Radio New Zealand, 2011a). The government's OEPPC has coordinated the Global Environment Facility-supported Pacific Adaptation to Climate Change project, aiming in particular to protect water resources in Majuro through a partnership with Majuro Water and Sewer Company, aa well as an ADB-funded community vulnerability survey of Rita district, Majuro, and topographical survey of Majuro to provide fine-scale predictions of flooding. The measures being put in place include the use of reverse osmosis water-making machines, composting toilets to reduce contamination of the Laura water lens (one of Majuro's main water sources), the installation of leak-detecting equipment and evaporation-reducing covers at Majuro's airport reservoir (another major water source), and the use of climate-simulation software developed by CLIMsystems Ltd. of New Zealand to plan for various climate scenarios. In 2012, the presidents of the Marshall Islands, the Federated States of Micronesia, and Palau signed a communiqué declaring their mutual intention to pursue debt-swap-for-adaptation opportunities.

Bibliography

ABC (2008). Marshall Islands in "Major Clean-Up." *Australian Broadcasting Corporation*, December 26. Available at http://www.abc.net.au/news/stories/2008/12/26/2455403.htm?site=news

Abo, T., Bender, B., Capelle, A., & DeBrum, T. (1976). *Marshallese-English Dictionary*. Honolulu: University of Hawaii Press.

Adger, W. N., & Barnett, J. (2005). Compensation for Climate Change Must Meet Needs. *Nature*, 436(21), Correspondence.

Adger, W. N., Barnett, J., & Ellemor, H. (2010). Unique and Valued Places. In S. H. Schneider, A. Rosencranz, M. D. Mastrandrea, & K. Kuntz-Duriseti (Eds.), *Climate Change Science and Policy* (pp. 131–138). Washington, DC: Island Press.

Alicke, M. D. (2000). Culpable Control and the Psychology of Blame. *Psychological Bulletin*, 126(4), 556–574.

Alkire, W. H. (1999). Cultural Ecology and Ecological Anthropology in Micronesia. In R. C. Kiste & M. Marshall (Eds.), *American Anthropology in Micronesia: An Assessment* (pp. 81–105). Honolulu: University of Hawaii Press.

Asian Development Bank (1997). Marshall Islands 1996 Economic Report. Manila: Asian Development Bank.

Australian Government (2011). Current and Future Climate of the Marshall Islands. International Climate Change Adaptation Initiative / Pacific Climate Change Science Program / Australian AID / Australian Bureau of Meteorology / Australian Department of Climate Change and Energy Efficiency / CSIRO / Marshall Islands National Weather Service. Available at http://www.cawcr.gov.au/projects/PCCSP/pdf/8_PCCSP_Marshall_Islands_8pp.pdf

Baer, H. A. (2011). The International Climate Justice Movement: A Comparison with the Australian Climate Movement. *The Australian Journal of Anthropology*, 22(2), 256–260.

Ballard, C. (2000). The Fire Next Time: The Conversion of the Huli Apocalypse. *Ethnohistory*, 47(1), 205–225.

Barker, H. M. (2004). *Bravo for the Marshallese: Regaining Control in a Post-Nuclear, Post-Colonial World*. Belmont, CA: Wadsworth.

Barker, H. M. (2008). The Inequities of Climate Change and the Small Island Experience. *Counterpunch*, November 4.

Barlett, P. E., & Stewart, B. (2009). Shifting the University: Faculty Engagement and Curriculum Change. In S. A. Crate & M. Nuttall (Eds.), *Anthropology & Climate Change: From Encounters to Actions* (pp. 356–369). Walnut Creek, CA: Left Coast Press.

Barnett, J. (2005). Titanic States? Impacts and Responses to Climate Change in the Pacific Islands. *Journal of International Affairs*, 59(1), 203–219.

Barnett, J., & Adger, W. N. (2003). Climate Dangers and Atoll Countries. *Climatic Change*, 61, 321–337.

Barnett, J., & Campbell, J. (2010). *Climate Change and Small Island States: Power, Knowledge and the South Pacific*. London: Earthscan.
Barnett, J., & Webber, M. (2010). Migration as Adaptation: Opportunities and Limits. In J. McAdam (Ed.), *Climate Change and Displacement: Multidisciplinary Perspectives* (pp. 37–55). Oxford, UK, and Portland, OR: Hart Publishing.
Bataua, B. (2009a). Only 50 Years Left for Kiribati? *Marshall Islands Journal*, January 23, p. 9.
Bataua, B. (2009b). Tong Looks for Land. *Marshall Islands Journal*, March 20, p. 9.
Batterbury, S. (2008). Anthropology and Global Warming: The Need for Environmental Engagement. *The Australian Journal of Anthropology*, 19, 62–68.
BBC (2008). Marshall Atolls Declare Emergency. *BBC News*. Available at http://news.bbc.co.uk/1/hi/world/asia-pacific/7799566.stm
BBC World Service Trust (2010). Africa Talks Climate Research Report: Executive Summary. *BBC World Service Trust and British Council*. Available at http://africatalksclimate.com/sites/default/files/01-Executive Summary.pdf
Beaglehole, E. (1941). The Polynesian Maori. In The Polynesian Society (Ed.), *Polynesian Anthropological Studies* (pp. 39–67). New Plymouth, New Zealand: Thomas Avery and Sons.
Beck, U. (1992). *Risk Society: Towards a New Modernity* (Transl. Mark Ritter). London: Sage.
Berkes, F., & Jolly, D. (2001). Adapting to Climate Change: Social-Ecological Resilience in a Canadian Western Arctic Community. *Conservation Ecology*, 5(2), 18.
Berrang-Ford, L., Ford, J. D., & Paterson, J. (2011). Are We Adapting to Climate Change ? *Global Environmental Change*, 21(1), 25–33.
Berreman, G. D. (1962). *Behind Many Masks: Ethnography and Impression Management in a Himalayan Village*. Ithaca, NY: Society for Applied Anthropology.
Besnier, N. (2004). Authority and Egalitarianism: Discourses of Leadership on Nukulaelae Atoll. In R. Feinberg & K. A. Watson-Gegeo (Eds.), *Leadership and Change in the Western Pacific: Essays Presented to Sir Raymond Firth on the Occasion of his Ninetieth Birthday* (pp. 93–128). Oxford, UK: Berg.
Bigler, C. (2007). I'm Not Ready to Live under Water! *Marshall Islands Journal*, July 27, p. 16.
Bloch, M. (2000). A Well-Disposed Social Anthropologist's Problems with Memes. In R. Aunger (Ed.), *Darwinizing Culture: The Status of Memetics as a Science* (pp. 189–203). Oxford, UK: Oxford University Press.
Bloch, M., & Parry, J. (1989). Introduction. In M. Bloch & J. Parry (Eds.), *Money and the Morality of Exchange*. Cambridge, UK: Cambridge University Press.
Boholm, Å. (1996). Risk Perception and Social Anthropology: Critique of Cultural Theory. *Ethnos*, 61(1), 64–84.
Bohren, L. (2009). Car Culture and Decision-Making: Choice and Climate Change. In S. A. Crate & M. Nuttall (Eds.), *Anthropology & Climate Change: From Encounters to Actions* (pp. 370–379). Walnut Creek, CA: Left Coast Press.
Booth, H. (1999). Pacific Island Suicide in Comparative Perspective. *Journal of the Biosocial Society*, 31, 433–448.
Borofsky, R. (1987). *Making History: Pukapukan and Anthropological Constructions of Knowledge*. Cambridge, UK: Cambridge University Press.
Bostrom, A., & Lashof, D. (2007). Weather or Climate Change? In S. C. Moser & L. Dilling (Eds.), *Creating a Climate for Change: Communicating Climate Change and Facilitating Social Change* (pp. 31–43). Cambridge, UK: Cambridge University Press.
Bourdieu, P. (1994). *Language and Symbolic Power* (Transl. Gino Raymond and Matthew Adamson). Cambridge, MA: Harvard University Press.

Boykoff, M. T., & Roberts, J. T. (2007). Media Coverage of Climate Change: Current Trends, Strengths, Weaknesses. Human Development Report Office Occasional Paper. Available at http://hdr.undp.org/en/reports/global/hdr2007–2008/papers/boykoff, maxwell and roberts, j. timmons.pdf

Bravo, M. T. (2009). Voices from the Sea Ice: the Reception of Climate Impact Narratives. *Journal of Historical Geography*, 35, 256–278.

Brenot, J., Bonnefous, S., & Marris, C. (1998). Testing the Cultural Theory of Risk in France. *Risk Analysis*, 18(6), 729–739.

Bridges, K. W., & McClatchey, W. C. (2009). Living on the Margin: Ethnoecological Insights from Marshall Islanders at Rongelap Atoll. *Global Environmental Change*, 19, 140–146.

Broad, K., & Orlove, B. (2007). Channeling Globality: The 1997–98 El Niño Climate Event in Peru. *American Ethnologist*, 34(2), 285–300.

Brugger, J. (2010). Why Americans Don't Believe in Climate Change. Paper presented at the Annual Meeting of the American Anthropological Association, New Orleans, November 17–21.

Brunton, R. (1980). Misconstrued Order in Melanesian Religion. *Man*, 15(1), 112–128.

Byg, A., & Salick, J. (2009). Local Perspectives on a Global Phenomenon—Climate Change in Eastern Tibetan Villages. *Global Environmental Change*, 19, 156–166.

Camino, L. A. (1994). Refugee Adolescents and Their Changing Identities. In L. A. Camino & R. M. Krulfeld (Eds.), *Reconstructing Lives, Recapturing Meaning: Refugee Identity, Gender, and Culture Change*. Basel: Gordon and Breach Science Publishers.

Campbell, J. (2010). Climate-Induced Community Relocation in the Pacific: The Meaning and Importance of Land. In J. McAdam (Ed.), *Climate Change and Displacement: Multidisciplinary Perspectives* (pp. 57–79). Oxford, UK, and Portland, OR: Hart Publishing.

Campbell, J. R., Goldsmith, M., & Koshy, K. (2005). Community Relocation as an Option for Adaptation to the Effects of Climate Change and Climate Variability in Pacific Island Countries (PICs). Final Report for APN Project 2005–14-NSY-Campbell. Kobe, Japan: Asia-Pacific Network for Global Change Research.

Carucci, L. M. (1984). Significance of Change or Change of Significance: A Consideration of Marshallese Personal Names. *Ethnology*, 23(2), 143–155.

Carucci, L. M. (1986). Sly Moves: Ritual Movements in Marshallese Culture. *Semiotica*, 62(1–2), 165–177.

Carucci, L.M., (1989). The Source of the Force in Marshallese Cosmology. In L. Lindstrom & G. White, eds. *The Pacific Theatre: Island Representations of World War II* (pp. 73–96). Honolulu: University of Hawaii Press.

Carucci, L. M. (1990). Negotiations of Violence in the Marshallese Household. *Pacific Studies*, 13(3), 93–113.

Carucci, L. M. (1993). Medical Magic and Medicinal Cure: Manipulating Meanings with Ease of Disease. *Cultural Anthropology*, 8(2), 157–168.

Carucci, L. M. (1997a). Irooj Ro Ad: Measures of Chiefly Ideology and Practice in the Marshall Islands. In G. M. White & L. Lindstrom (Eds.), *Chiefs Today: Traditional Pacific Leadership and the Postcolonial State* (pp. 197–210). Stanford, CA: Stanford University Press.

Carucci, L. M. (1997b). *Nuclear Nativity: Rituals of Renewal and Empowerment in the Marshall Islands*. DeKalb, IL: Northern Illinois University Press.

Carucci, L. M. (1998). Working Wrongly and Seeking the Straight: Remedial Remedies on Enewetak Atoll. *Pacific Studies*, 21(3), 1–27.

Carucci, L. M. (2003). The Church as Embodiment and Expression of Community on Ujelang and Enewetak, Marshall Islands. *Pacific Studies*, 26(3–4), 55–78.

Cass, L. R. (2007). Measuring the Domestic Salience of International Environmental Norms: Climate Change Norms in American, German, and British Climate Policy Debates. In M. E. Pettenger (Ed.), *The Social Construction of Climate Change: Power, Knowledge, Norms, Discourses* (pp. 23–50). Aldershot, UK: Ashgate Publishing Limited.

Chakrabarty, D. (2009). The Climate of History: Four Theses. *Critical Inquiry*, 35, 197–222.

Chamisso, A. von (1986[1821]). *A Voyage around the World with the Romanzov Exploring Expedition in the Years 1815–1818 in the Brig Rurick, Captain Otto von Kotzebue* (transl. H. Kratz). Honolulu: University of Hawaii Press.

Chapman, P. (2012). Entire Nation of Kiribati to Be Relocated over Rising Sea Level Threat. *The Telegraph*, March 7, 2012. Available at http://www.telegraph.co.uk/news/worldnews/australiaandthepacific/kiribati/9127576/Entire-nation-of-Kiribati-to-be-relocated-over-rising-sea-level-threat.html

Chess, C., & Johnson, B. B. (2007). Information Is Not Enough. In S. C. Moser & L. Dilling (Eds.), *Creating a Climate for Change: Communicating Climate Change and Facilitating Social Change*. Cambridge, UK: Cambridge University Press.

Clifford, J. (1988). *The Predicament of Culture*. Cambridge, MA: Harvard University Press.

Condry, E. (1976). The Impossibility of Solving the Highland Problem. *Journal of the Anthropological Society of Oxford*, 7(3), 138–149.

Connell, J. (2003). Losing Ground? Tuvalu, the Greenhouse Effect and the Garbage Can. *Asia Pacific Viewpoint*, 44(2), 89–107.

Crate, S. A. (2008). Gone the Bull of Winter? Grappling with the Cultural Implications of and Anthropology's Role(s) in Global Climate Change. *Current Anthropology*, 49(4), 569–595.

Crate, S. A. (2011). Climate and Culture: Anthropology in the Era of Contemporary Climate Change. *Annual Review of Anthropology*, 40(1), 175–194.

Crate, S. A., & Nuttall, M. (2009a). Epilogue: Anthropology, Science, and Climate Change Policy. In S. A. Crate & M. Nuttall (Eds.), *Anthropology & Climate Change: From Encounters to Actions* (pp. 394–400). Walnut Creek, CA: Left Coast Press.

Crate, S. A., & Nuttall, M. (Eds.). (2009b). *Anthropology & Climate Change: From Encounters to Actions*. Walnut Creek, CA: Left Coast Press.

Crate, S. A., & Nuttall, M. (2009c). Introduction: Anthropology and Climate Change. In S. A. Crate & M. Nuttall (Eds.), *Anthropology & Climate Change: From Encounters to Actions* (pp. 9–36). Walnut Creek, CA: Left Coast Press.

Crawford, R. (1987). Cultural Influences on Prevention and the Emergence of a New Health Consciousness. In N. D. Weinstein (Ed.), *Taking Care: Understanding and Encouraging Self-Protective Behavior*. Cambridge, UK: Cambridge University Press.

Crocombe, R. (1994). Cultural Policies in the Pacific Islands. In L. Lindstrom & G. M. White (Eds.), *Culture—Kastom—Tradition: Developing a Cultural Policy in Melanesia*. Suva, Fiji: Institute of Pacific Studies of the University of the South Pacific.

Cruikshank, J. (2005). *Do Glaciers Listen? Local Knowledge, Colonial Encounters, & Social Imagination*. Vancouver: UBC Press.

Daniels, S., & Endfield, G. H. (2008). Narratives of Climate Change: Introduction. *Journal of Historical Geography*, 35, 215–222.

Dawkins, R. (1995). *River Out of Eden: A Darwinian View of Life*. Basic Books.

Dawkins, R. (2006[1976]). *The Selfish Gene*. Oxford, UK: Oxford University Press.

DeBrum, J. (2010). Kids Enjoy OEPPC Energy Fair. *Marshall Islands Journal*, March 19, p. 10.

Degawan, M. (2008). Mitigating the Impacts of Climate Change: Solutions or Additional Threats? *Indigenous Affairs*, 1–2, 52–59.

Deunert, B., et al. (1999). *Anthropological Survey of Jaluit Atoll: Terrestrial and Underwater Reconnaissance Surveys and Oral History Recordings*. Majuro, Marshall Islands: Republic of the Marshall Islands Historic Preservation Office.

Dibblin, J. (1988). *Day of Two Suns: US Nuclear Testing and the Pacific Islanders*. London: Virago.

Donner, S. D. (2007). Domain of the Gods: An Editorial Essay. *Climatic Change*, 85, 231–236.

Donner, S. D. (2011). Making the Climate a Part of the Human World. *Bulletin of the American Meteorological Society*, 92(10), 1297–1302.

Douglas, M. (1970). *Natural Symbols: Explorations in Cosmology*. London: Barrie and Jenkins.

Douglas, M. (Ed.) (1982a). *Essays in the Sociology of Perception*. London: Routledge & Kegan Paul.

Douglas, M. (1982b). Introduction to Grid/Group Analysis. In M. Douglas (Ed.), *Essays in the Sociology of Perception* (pp. 1–8). London: Routledge & Kegan Paul.

Douglas, M. (1992). *Risk and Blame: Essays in Cultural Theory*. New York: Routledge.

Douglas, M. (2002 [1966]). *Purity and Danger: An Analysis of Concepts of Pollution and Taboo*. New York: Routledge.

Douglas, M., & Wildavsky, A. B. (1982). *Risk and Culture: An Essay on the Selection of Technical and Environmental Dangers*. Berkeley: University of California Press.

Easterbrook, G. (1996). *A Moment on the Earth: The Coming Age of Environmental Optimism*. London and New York: Penguin.

Edelstein, E. L., Nathanson, D. L., & Stone, A. M. (Eds.). (1989). *Denial: A Clarification of Concepts and Research*. New York: Plenum Press.

Einhorn, H. J., & Hogarth, R. M. (1986). Judging Probable Cause. *Psychological Bulletin*, 99, 3–19.

Eisenbruch, M. (1990). Cultural Bereavement and Homesickness. In S. Fisher & C. L. Cooper (Eds.), *On the Move: The Psychology of Change and Transition*. Chichester, UK: John Wiley & Sons.

Erdland, A. (1961[1914]). *The Marshall Islanders: Life and Customs, Thoughts and Religion of a South Seas People* (Translated from the German by Richard Neuse). Ethnological Monographs 2(1). Human Relations Area files. New Haven, CT: Anthropos Bibliothek.

Errington, F., & Gewertz, D. (1986). The Confluence of Powers: Entropy and Importation among the Chambri. *Oceania*, 57(2), 99–113.

Escobar, A. (1999). After Nature: Steps to an Antiessentialist Political Ecology. *Current Anthropology*, 40(1), 1–30.

Farbotko, C. (2005). Tuvalu and Climate Change: Constructions of Environmental Displacement in the Sydney Morning Herald. *Geografiska Annaler, Series B: Human Geography*, 87(4), 279–293.

Farbotko, C. (2010). Wishful Sinking: Disappearing Islands, Climate Refugees and Cosmopolitan Experimentation. *Asia Pacific Viewpoint*, 51(1), 47–60.

Farbotko, C., & Lazrus, H. (2012). The First Climate Refugees? Contesting Global Narratives of Climate Change in Tuvalu. *Global Environmental Change*, 22(2), 382–390.

Feinberg, M., & Willer, R. (2011). Apocalypse Soon? Dire Messages Reduce Belief in Global Warming by Contradicting Just-World Beliefs. *Psychological Science*, 22(1), 34–38.

Feygina, I., Jost, J. T., & Goldsmith, R. E. (2010). System Justification, the Denial of Global Warming, and the Possibility of "System-Sanctioned Change." *Personality and Social Psychology Bulletin*, 36(3), 326–338.

Finan, T. J. (2007). Is "Official" Anthropology Ready for Climate Change? *Anthropology News*, 48(9), 10–11.

Firth, R. (1963[1936]). *We, The Tikopia*. Boston: Beacon Press.

Flinn, J. (1990). We Still Have Our Customs: Being Pulapese in Truk. In J. Linnekin & L. Poyer (Eds.), *Cultural Identity and Ethnicity in the Pacific* (pp. 103–126). Honolulu: University of Hawaii Press.

Floyd, D. L., Prentice-Dunn, S., & Rogers, R. W. (2000). A Meta-Analysis of Research on Protection Motivation Theory. *Journal of Applied and Social Psychology*, 30, 407–429.

Ford, M. (2012). Shoreline Changes on an Urban Atoll in the Central Pacific Ocean: Majuro Atoll, Marshall Islands. *Journal of Coastal Research*, 279(1), 11–22.

Fosberg, F. R. (1955). Northern Marshall Islands Expedition, 1951–1952. Narrative. *Atoll Research Bulletin*, 38.

Fox, S. (2002). These Are Things That Are Really Happening: Inuit Perspectives on the Evidence and Impacts of Climate Change in Nunavut. In I. Krupnik & D. Jolly (Eds.), *The Earth Is Faster Now: Indigenous Observations of Arctic Environmental Change* (pp. 12–53). Washington, DC: Arctic Research Consortium of the United States in cooperation with the Arctic Studies Center, Smithsonian Institution.

Frederick, S., Loewenstein, G., & O'Donoghue, T. (2002). Time Discounting and Time Preference: A Critical Review. *Journal of Economic Literature*, 40, 351–401.

Freedom Report (2009). International Religious Freedom Report 2009. Bureau of Democracy, Human Rights, and Labor. Available at http://www.state.gov/g/drl/rls/irf/2009/127278.htm

Garcia-Alix, L. (2008). The United Nations Permanent Forum on Indigenous Issues Discusses Climate Change. *Indigenous Affairs*, 1–2, 16–23.

Gellner, E. (1970). Concepts and Society. In B. R. Wilson (Ed.), *Rationality* (pp. 18–49). New York: Harper & Row.

Gellner, E. (1985). *Relativism and the Social Sciences*. Cambridge, UK: Cambridge University Press.

Giddens, A. (1999). Risk and Responsibility. *The Modern Law Review*, 62(1), 1–10.

Giddens, A. (2002). *Runaway World: How Globalisation is Reshaping Our Lives*. London: Profile Books.

Giddens, A. (2009). *The Politics of Climate Change*. Cambridge, UK: Polity Press.

Goffman, E. (1959). *The Presentation of Self in Everyday Life*. New York: Doubleday.

Gold, A. G. (1998). Sin and Rain: Moral Ecology in Rural North India. In L. E. Nelson (Ed.), *Purifying the Earthly Body of God: Religion and Ecology in Hindu India* (pp. 165–195). Albany: State University of New York Press.

Gombrich, R. F. (1971). *Precept and Practice: Traditional Buddhism in the Rural Highlands of Ceylon*. Oxford, UK: Clarendon Press.

Gore, A. 2006. *An Inconvenient Truth*. Film directed by Davis Guggenheim. Hollywood: Paramount Pictures.

Graveling, E. (2010). "That is Not Religion, That Is the Gods": Ways of Conceiving Religious Practices In Rural Ghana. *Culture and Religion*, 11(1), 31–50.

Grothmann, T., & Patt, A. G. (2005). Adaptive Capacity and Human Cognition: The Process of Individual Adaptation to Climate Change. *Global Environmental Change Part A*, 15(3), 199–213.

Grothmann, T., & Reusswig, F. (2006). People at Risk of Flooding: Why Some Residents Take Precautionary Action While Others Do Not. *Natural Hazards*, 38, 101–120.

Guilcher, A. (1988). *Coral reef geomorphology*. London: John Wiley & Sons.

Haddon, A. C., & J. Hornell (1975). *Canoes of Oceania*. Honolulu: Bishop Museum.

Harrell-Bond, B. (1999). The Experience of Refugees as Recipients of Aid. In A. Ager (Ed.), *Refugees: Perspectives on the Experience of Forced Migration*. London: Pinter.

Hassan, F. A. (2009). Human Agency, Climate Change, and Culture: An Archaeological Perspective. In S. A. Crate & M. Nuttall (Eds.), *Anthropology & Climate Change: From Encounters to Actions* (pp. 39–69). Walnut Creek, CA: Left Coast Press.

Hassol, S. J. (2008). Improving How Scientists Communicate about Climate Change. *Weekly Journal of the American Geophysical Union*, 89, 106–107.

Hau'ofa, E. (1993). Our Sea of Islands. In E. Waddell, V. Naidu, & E. Hau'ofa (Eds.), *A New Oceania: Rediscovering our Sea of Islands* (pp. 2–16). Suva, Fiji: School of Social and Economic Development, University of the South Pacific, and Beake House.

Heine, C. (2000). The Bible Code and the Year 2000. *Marshall Islands Journal*, January 7, p. 15.

Heine, H. (2010). Science Affair. *Marshall Islands Journal*, July 23, p. 12.

Heine, H. C., Maddison, M. L., & Rechebei, E. D. (2009). Education for Sustainable Development in the Pacific: A Mapping Analysis of the Republic of the Marshall Islands. Honolulu: Pacific Resources for Education and Learning.

Henry, D. (2010). RMI under Threat. *The Marshall Islands Journal*, March 2, p. 10.

Hezel, F. X. (1983). *The First Taint of Civilization: A History of the Caroline and Marshall Islands in Pre-Colonial Days, 1521–1885*. Honolulu: University of Hawaii Press.

Hezel, F. X, (2009a). High Water in the Low Atolls. *Micronesian Counselor*. Available at http://www.micsem.org/pubs/counselor/frames/highwaterfr.htm

Hezel, F. X. (2009b). A Teacher's Tale. *Micronesian Counselor*, 78, 2–22.

Hitchcock, R. K. (2009). From Local to Global: Perceptions and Realities of Environmental Change among Kalahari San. In S. A. Crate & M. Nuttall (Eds.), *Anthropology & Climate Change: From Encounters to Actions* (pp. 250–261). Walnut Creek, CA: Left Coast Press.

Holthus, P., Crawford, M., Makroro, C., & Sullivan, S. (1992). Vulnerability Assessment of Accelerated Sea Level Rise: Case Study: Majuro Atoll, Marshall Islands. Washington, DC: United States National Oceanic and Atmospheric Administration (NOAA) / South Pacific Regional Environment Programme.

Howard, A. (1990). Cultural Paradigms, History, and the Search for Identity in Oceania. In J. Linnekin & L. Poyer (Eds.), *Cultural Identity and Ethnicity in the Pacific*. Honolulu: University of Hawaii Press.

Hsu, C. (2000). Chinese Attitudes toward Climate. In R. J. McIntosh, J. A. Tainter, & S. K. McIntosh (Eds.), *The Way the Wind Blows: Climate, History, and Human Action* (pp. 209–222). New York: Columbia University Press.

Huber, T., & Pedersen, P. (1997). Meteorological Knowledge and Environmental Ideas in Traditional and Modern Societies: The Case of Tibet. *Journal of the Royal Anthropological Institute*, 3(3), 577–597.

Hulme, M. (2009). *Why We Disagree about Climate Change: Understanding Controversy, Inaction and Opportunity*. Cambridge, UK: Cambridge University Press.

Hulme, M. (2010a). So What Does Climate Change Mean for You? Presentation at Climate Week, Environmental Change Institute, Oxford University, October 14, 2010.

Hulme, M. (2010b). Learning to Live with Recreated Climates. *Nature and Culture*, 5(2), 117–122.
Hulme, M. (2010c). Cosmopolitan Climates: Hybridity, Foresight and Meaning. Theory, *Culture & Society*, 27(2–3), 267–276.
Huntington, H. P., Fox, S., Berkes, F., & Krupnik, I. (2005). The Changing Arctic: Indigenous Perspectives. In ACIA (Ed.), *Arctic Climate Impact Assessment* (pp. 61–98). Cambridge, UK: Cambridge University Press.
Höche, S. (2009). *Cognate Object Constructions in English: a Cognitive-Linguistic Account*. Tübingen, Germany: Gunter Narr Verlag.
Ingold, T. (2008[1993]). Globes and Spheres: The Topology of Environmentalism. In M. R. Dove & C. Carpenter (Eds.), *Environmental Anthropology: A Historical Reader* (pp. 462–469). Malden, MA: Blackwell Publishing.
IPCC (2007a). *Climate Change 2007: The Physical Science Basis. Contribution of Working Group I to the Fourth Assessment Report of the Intergovernmental Panel on Climate Change*. (S. Solomon et al., Eds.). Cambridge, UK: Cambridge University Press.
IPCC (2007b). *Climate Change 2007: Impacts, Adaptation and Vulnerability. Contribution of Working Group II to the Fourth Assessment Report of the Intergovernmental Panel on Climate Change*. (M. L. Parry, et al., Eds.). Cambridge, UK: Cambridge University Press.
Jack, A. I., & Roepstorff, A. (2003). Why Trust the Subject? *Journal of Consciousness Studies*, 10(9–10), v–xx.
Jacka, J. (2009). Global Averages, Local Extremes: The Subtleties and Complexities of Climate Change in Papua New Guinea. In S. A. Crate & M. Nuttall (Eds.), *Anthropology & Climate Change: From Encounters to Actions* (pp. 197–208). Walnut Creek, CA: Left Coast Press.
Jackson, C. (2010). Staying Afloat in Paradise: Reporting Climate Change in the Pacific. Fellowship Paper, Reuters Institute for the Study of Journalism, University of Oxford.
Jackson, J. (2006). Introducing Fear of Crime to Risk Research. *Risk Analysis*, 26(1), 253–264.
Jackson, J. E. (1994). Becoming Indians: The Politics of Tukanoan Ethnicity. In A. Roosevelt (Ed.), *Amazonian Indians from Prehistory to the Present*. Tucson: University of Arizona Press.
Jacobs, R. E. (2005). Treading Deep Waters: Substantive Law Issues in Tuvalu's Threat to Sue the United States in the International Court of Justice. *Pacific Rim Law and Policy Journal*, 14(1), 103–128.
Jamieson, D. (2007). The Moral and Political Challenges of Climate Change. In S. C. Moser & L. Dilling (Eds.), *Creating a Climate for Change: Communicating Climate Change and Facilitating Social Change* (pp. 475–482). Cambridge, UK: Cambridge University Press.
Jasanoff, S. (2005). *Designs on Nature: Science and Democracy in Europe and the United States*. Princeton, NJ, & Oxford, UK: Princeton University Press.
Jasanoff, S. (2010). A New Climate for Society. *Theory, Culture & Society*, 27(2–3), 233–253.
Jasanoff, S., & Wynne, B. (1998). Science and Decision Making. In S. Rayner & E. Malone (Eds.), *Human Choice and Climate Change: Volume 1. The Societal Framework* (pp. 1–87). Columbus, OH: Battelle Press.
Jennaway, M. (2008). Apocalypse on You! Millenarian Frenzy in Debates on Global Warming. *The Australian Journal of Anthropology*, 19(1), 68–73.
Jensen, A. (2012). *Rewriting Her Story*. Available at http://pacific.350.org/rewriting-her-story
Johnson, D., & Levin, S. (2009). The Tragedy of Cognition: Psychological Biases and Environmental Inaction. *Current Science*, 97(11), 1593–1603.

Johnson, G. (2004). Society at War: Special Journal Report. *Marshall Islands Journal*, January 23, pp. 14–15.
Johnson, G. (2008). Weather Event Shows Vulnerability of Atolls. *Marshall Islands Journal*, December 12, p. 4.
Johnson, G. (2009). Ene Ko Retta Renaj Ibwiji. *Marshall Islands Journal*, October 30, p. 13.
Jolly, D., Berkes, F., Castleden, J., Nichols, T., & the Community of Sachs Harbour. (2002). We Can't Predict the Weather Like We Used to: Inuvialuit Observations of Climate Change, Sachs Harbour, Western Canadian Arctic. In I. Krupnik & D. Jolly (Eds.), *The Earth is Faster Now: Indigenous Observations of Arctic Environmental Change* (pp. 92–125). Washington, DC: Arctic Research Consortium of the United States in cooperation with the Arctic Studies Center, Smithsonian Institution.
Jorgensen, D. (1981). Taro and Arrows: Order, Entropy, and Religion among the Telefolmin of Papua New Guinea. Doctoral dissertation, Department of Anthropology, University of British Columbia.
Jormelu, H. (2007). Kajin eo ad ej manit eo an ri-Majol. *Marshall Islands Journal*, March 30, p. 11.
Jormelu, K., Hagberg, E., & Goreau, T. J. (2010). Shore protection in the Republic of the Marshall Islands: Pilot Project Report. Report to the Federal Republic of Germany Federal Foreign Office Task Force for Humanitarian Aid.
Journal (2000). Beautiful Picture of Island Life. *Marshall Islands Journal*, April 21, p. 7.
Journal (2002). Drowning in Emotion. *Marshall Islands Journal*, April 12, p. 22.
Journal (2003a). Protestors Demand Delay on Compact. *Marshall Islands Journal*, August 8, p. 3.
Journal (2003b). Shocking Figures on Abuse of RMI Women. *Marshall Islands Journal*, October 3, p. 3.
Journal (2003c). Who Says Bus Stops Can't Be Beautiful? *Marshall Islands Journal*, November 14, p. 18.
Journal (2004a). Women "Barred" from Being Alabs. *Marshall Islands Journal*, January 30, p. 6.
Journal (2004b). Lymman Langijota: Our Special Force of One. *Marshall Islands Journal*, May 21, p. 8.
Journal (2005a). Future of RMI Is in the Ocean. *Marshall Islands Journal*, April 15, p. 9.
Journal (2005b). TRC: Women Can Hold Title of Alab. *Marshall Islands Journal*, April 22, p. 1.
Journal (2007a). Students Jump In for Their Islands. *Marshall Islands Journal*, June 8, p. 5.
Journal (2007b). Outer Islands Are Their Inner Dream. *Marshall Islands Journal*, November 30, p. 14.
Journal (2008a). Protest over New Dock Rule. *Marshall Islands Journal*, February 8, p. 2.
Journal (2008b). Wazzup with the Global Glow-Ball? *Marshall Islands Journal*, April 18, p. 14.
Journal (2008c). The Wavy Secret. *Marshall Islands Journal*, December 19, p. 7.
Journal (2009a). Emergency Set for Wave Damage. *Marshall Islands Journal*, January 2, 2009, p. 2.
Journal (2009b). Benedict Invites You to Global Day of Action. *Marshall Islands Journal*, October 23, p. 10.
Journal (2009c). Option for Sea Level Rise: Relocate. *Marshall Islands Journal*, October 23, pp. 1–2.

Journal (2010a). Highest Sea Level Ever in September. *Marshall Islands Journal*, November 26, p. 4.
Journal (2010b). "Ensuring Our Survival": Silk on the Risks of Climate Change. *Marshall Islands Journal*, February 5, p. 5.
Journal (2010c). Tastes Great, Is Great: It's Bōb. *Marshall Islands Journal*, April 30, pp. 12–13.
Journal (2010d). Ruben in S. Korea on Environment. *Marshall Islands Journal*, November 26, p. 16.
Journal (2010e). KALGov Gets Energy Grant. *Marshall Islands Journal*, April 2, p. 15.
Journal (2010f). Bōb Buddies. *Marshall Islands Journal*, May 7, p. 13.
Journal (2010g). Boy Scouts Learn Good Life Lessons. *Marshall Islands Journal*, July 2, p. 17.
Journal (2010h). Quote of the Week. *Marshall Islands Journal*, July 23, p. 7.
Journal (2010i). March against Abuse. *Marshall Islands Journal*, August 27, pp. 1–2.
Journal (2010j). Dealing with Coastal Flooding. *Marshall Islands Journal*, October 1, p. 10.
Journal (2010k). Ratak Aims for $51m. *Marshall Islands Journal*, December 31, pp. 1, 3.
Journal (2011a). Flooding Forecasted. *Marshall Islands Journal*, February 18, p. 3.
Journal (2011b). Things Are Heating Up in Majuro Atoll. *Marshall Islands Journal*, December 23, p. 22.
Journal (2011c). Enjoy the Rain while It Lasts. *Marshall Islands Journal*, December 23, p. 22.
Journal (2011d). USP Science Camp a Blast. *Marshall Islands Journal*, July 22, p. 23.
Journal (2011e). Brenson on Global Issue. *Marshall Islands Journal*, January 14, p. 6.
Journal (2011f). Bungitak: "Need to Protect Our Shoreline." *Marshall Islands Journal*, July 1, p. 14.
Journal (2012a). "Sea Level Rise Is Eating Up Our Land." *Marshall Islands Journal*, June 8, pp. 1, 3.
Journal (2012b). Debt Swap Is Latest Response to Climate Issue. *Marshall Islands Journal*, July 6, p. 4.
Juumemmej (2006). Juumemmej: Republic of the Marshall Islands Social and Economic Report 2005. Pacific Studies Series. Manila: Asian Development Bank.
Kabua, A. (1993). *Customary Titles and Inherent Rights: A General Guideline in Brief.* Majuro: Republic of the Marshall Islands.
Kabua, C. (2010). My Five Senses Can Perceive Climate Change. *Marshall Islands Journal*, January 15, p. 11.
Kabua, M. (2008). Mike Kabua: erub ke mantin Majol rainin? *Marshall Islands Journal*, January 11, pp. 12–13.
Kahan, D. M., Braman, D., Slovic, P., Gastil, J., & Cohen, G. (2007). The Second National Risk and Culture Study: Making Sense of—and Making Progress In-the American Culture War of Fact. The Cultural Cognition Project at Yale Law School.
Kahn, E. J. (1966). *A Reporter in Micronesia.* New York: W.W. Norton & Company.
Kasperson, R. E., & Kasperson, J. X. (2005[1991]). Hidden Hazards. In J. X. Kasperson & R. E. Kasperson (Eds.), *The Social Contours of Risk: Publics, Risk Communication and the Amplification of Risk* (Vol. 1, pp. 115–132). London: Earthscan.

Keane, B. (2009). Kiribati and Tuvalu Climate Change Strategy: Total Evacuation. Available at http://www.crikey.com.au/2009/07/27/kiribati-and-tuvulu-plan-climate-change-strategy-total-evacuation/

Keesing, R. (1994). Responsibilities of Long-Term Research. In L. Lindstrom & G. M. White (Eds.), *Culture—Kastom—Tradition: Developing a Cultural Policy in Melanesia*. Fiji: Institute of Pacific Studies of the University of the South Pacific.

Keesing, R. (2000). Creating the Past: Custom and Identity in the Contemporary Pacific. In D. L. Hanlon & G. M. White (Eds.), *Voyaging Through the Contemporary Pacific* (pp. 231–251). Lanham, MD: Rowman and Littlefield.

Kelin, D. A. (2003). *Marshall Islands Legends and Stories*. Honolulu: Bess Press.

Kempf, W. (2009). A Sea of Environmental Refugees? Oceania in an Age of Climate Change. In E. Hermann, K. Klenke, & M. Dickhardt (Eds.), *Form, Macht, Differenz: Motive und Felder Ethnologischen Forschens* (pp. 191–205). Gottingen, Germany: Universitatsverlag Gottingen.

Kench, P., et al. (2011). Improving Understanding of Local-Scale Vulnerability in Atoll Island Countries: Developing Capacity to Improve In-Country Approaches and Research. Final Report to the Asia-Pacific Network for Global Change Research. Available at http://www.apn-gcr.org/resources/archive/files/c54841d9632f99c90e37bb8e8fc208e6.pdf

Kirsch, S. (2001). Lost Worlds: Environmental Disaster, "Culture Loss," and the Law. *Current Anthropology*, 42(2), 167–198.

Kiste, R. C. (1974). *The Bikinians: A Study in Forced Migration*. Menlo Park, CA: Cummings Publishing Company.

Kiste, R. C. (1977). Relocation of the Bikini Marshallese. In M. D. Lieber (Ed.), *Exiles and Migrants in Oceania* (pp. 81–120). Honolulu: University of Hawaii Press.

Kluge, P. F. (1993). *The Edge of Paradise: America in Micronesia*. Honolulu: University of Hawaii Press.

Knudsen, K. E. (1977). Sydney Island, Titiana, and Kamaleai: Southern Gilbertese in the Phoenix and Solomon Islands. In M. D. Lieber (Ed.), *Exiles and Migrants in Oceania* (pp. 195–242). Honolulu: University of Hawaii Press.

Kotzebue, O. von (1821). *A Voyage of Discovery into the South Sea and Beering's Straits, for the Purpose of Exploring a North-East Passage, Undertaken in the Years 1815–1818, at the Expense of His Highness the Chancellor of the Empire, Count Romanzoff, in the Ship Rurick, under the Command of the Lieutenant in the Russian Imperial Navy, Otto von Kotzebue*, Volume II (Translated by H. E. Lloyd). London: Longman, Hurst, Rees, Orme and Brown.

Kramer, A., & Nevermann, H. (1938). *Ralik—Ratak (Marshall Islands)* (Transl. C. Brant and J.M. Armstrong). Hamburg: Friederichsen, De Gruyter and Co.

Kroemker, D., & Mosler, H.-J. (2002). Human Vulnerability—Factors Influencing the Implementation of Prevention and Protection Measures: An Agent Based Approach. In K. Steininger & H. Weck-Hannemann (Eds.), *Global Environmental Change in Alpine Regions: Impact, Recognition, Adaptation, and Mitigation* (pp. 95–114). Cheltenham: Edward Elgar.

Krupnik, I., & Jolly, D. (Eds.). (2002). *The Earth Is Faster Now: Indigenous Observations of Arctic Environmental Change*. Washington, DC: Arctic Research Consortium of the United States in cooperation with the Arctic Studies Center, Smithsonian Institution.

Kuruppu, N., & Liverman, D. (2011). Mental Preparation for Climate Adaptation: The Role of Cognition and Culture in Enhancing Adaptive Capacity of Water Management in Kiribati. *Global Environmental Change*, 21, 657–669.

Lahsen, M. (2004). Transnational Locals: Brazilian Experiences of the Climate Regime. In S. Jasanoff & M. L. Martello (Eds.), *Earthly Politics: Local and*

Global in Environmental Governance (pp. 151–172). Cambridge, MA: MIT Press.

Lahsen, M. (2007a). Anthropology and the Trouble of Risk Society. *Anthropology News*, 48(9), 9–10.

Lahsen, M. (2007b). Trust through Participation? Problems of Knowledge in Climate Decision-Making. In M. E. Pettenger (Ed.), *The Social Construction of Climate Change: Power, Knowledge, Norms, Discourses* (pp. 173–196). Aldershot, UK: Ashgate Publishing.

Lahsen, M. (2010). The Social Status of Climate Change Knowledge: An Editorial Essay. *Wiley Interdisciplinary Reviews: Climate Change*, 1(2), 162–171.

Lanman, J. A. (2007). How "Natives" Don't Think: The Apotheosis of Overinterpretation. In H Whitehouse & J. Laidlaw (Eds.), *Religion, Anthropology, and Cognitive Science* (pp. 105–132). Durham, NC: Carolina Academic Press.

Lanman, J. A . (2008). In Defence of "Belief:" A Cognitive Response to Behaviourism, Eliminativism, and Social Constructivism. *Issues in Ethnology and Anthropology*, 3, 49–62.

Larson, E. H. (1977). Tikopia in the Russell Islands. In M. D. Lieber (Ed.), *Exiles and Migrants in Oceania*. Honolulu: University of Hawaii Press.

Larson, J. C. (2000). Racing the Rising Tide: Legal Options for the Marshall Islands. *Michigan Journal of International Law*, 21, 485–522.

Latour, B. (1993). *We Have Never Been Modern* (transl. Catherine Porter). Cambridge, MA: Harvard University Press.

Latour, B. (2004). *Politics of Nature: How to Bring the Sciences into Democracy* (transl. Catherine Porter). Cambridge, MA: Harvard University Press.

Lawson, S. (1990). The Myth of Cultural Homogeneity and Its Implications for Chiefly Power and Politics in Fiji. *Comparative Studies in Society and History*, 32(4), 795–821.

Lawson, S. (1993). The Politics of Tradition: Problems for Political Legitimacy and Democracy in the South Pacific. *Pacific Studies*, 16(2), 1–29.

Lazrus, H. (2009). The Governance of Vulnerability: Climate Change and Agency in Tuvalu, South Pacific. In S. A. Crate & M. Nuttall (Eds.), *Anthropology & Climate Change: From Encounters to Actions* (pp. 240–249). Walnut Creek, CA: Left Coast Press.

Leary, M. R. (1995). *Self-Presentation: Impression Management and Interpersonal Behavior.* Madison, WI: Brown and Benchmark.

Leary, M. R., & Kowalski, R. M. (1990). Impression Management: A Literature Review and Two-Component Model. *Psychological Bulletin*, 107(1), 34–47.

Leduc, T. B. (2007). Sila Dialogues on Climate Change: Inuit Wisdom for a Cross-Cultural Interdisciplinarity. *Climatic Change*, 85, 237–250.

Leduc, T. B. (2011). *Climate, Culture, Change: Inuit and Western Dialogues with a Warming North.* Ottawa: University of Ottawa Press.

Lee, T. (1990). Moving House and Home. In S. Fisher & C. L. Cooper (Eds.), *On the Move: The Psychology of Change and Transition.* Chichester: John Wiley & Sons.

Lehman, D. R., & Taylor, S. E. (1987). Date with an Earthquake: Coping with a Probable, Unpredictable Disaster. *Personality and Social Psychology Bulletin*, 13(4), 546–555.

Lindisfarne, N. (2010). Cochabamba and Climate Anthropology. *Anthropology Today*, 26(4), 1–3.

Linnekin, J. S. (1983). Defining Tradition: Variations on the Hawaiian identity. *American Ethnologist*, 10, 241–252.

Lipset, D. (2011). The Tides: Masculinity and Climate Change in Coastal Papua New Guinea. *Journal of the Royal Anthropological Institute*, 17(1), 20–43.

Loeak, C. (2012). National Statement at the United Nations Conference on Sustainable Development—Rio+20. *Marshall Islands Journal*, June 29, p. 11.

Lord, C. G., Ross, L., & Lepper, M. R. (1979). Biased Assimilation and Attitude Polarization: The Effects of Prior Theories on Subsequently Considered Evidence. *Journal of Personality and Social Psychology*, 37, 2098–2109.

Lynas, M. (2004). *High Tide: How Climate Crisis Is Engulfing Our Planet*. London: Harper Perennial.

Lyon-Callo, V. (2000). Medicalizing Homelessness: The Production of Self-Blame and Self-Governing within Homeless Shelters. *Medical Anthropology Quarterly*, 14(3), 328–45.

Macnaghten, P., & Urry, J. (1995). Towards a Sociology of Nature. *Sociology*, 29(2), 203–220.

Maddux, J. E., & Rogers, R. W. (1983). Protection Motivation Theory and Self-Efficacy: A Revised Theory of Fear Appeals and Attitude Change. *Journal of Experimental Social Psychology*, 19, 469–479.

Madsen, A. (2006). Self-Made Artists Protect Our Culture. *Marshall Islands Journal*, October 6, p. 9.

Magistro, J., & Roncoli, C. (2001). Introduction: Anthropological Perspectives and Policy Implications of Climate Change Research. *Climate Research*, 19, 91–96.

Mail Foreign Service (2010). "The Planet Won't Be Destroyed by Global Warming because God Promised Noah," Says Politician Bidding to Chair U.S. Energy Committee. Dailymail.co.uk, 10 November 2010. Available at http://www.dailymail.co.uk/news/article-1328366/

Marino, E., & Schweitzer, P. (2009). Talking and Not Talking about Climate Change in Northwestern Alaska. In S. A. Crate & M. Nuttall (Eds.), *Anthropology & Climate Change: From Encounters to Actions* (pp. 209–217). Walnut Creek, CA: Left Coast Press.

Marks, K. (2008). Paradise Lost: Climate Change Forces South Sea Islanders to Seek Sanctuary Abroad. *The Independent*. Available at http://www.independent.co.uk/news/world/australasia/paradise-lost-climate-change-forces-south-sea-islanders-to-seek-sanctuary-abroad-841409.html

Marx, S., et al. (2007). Communication and Mental Processes: Experiential and Analytic Processing of Uncertain Climate Information. *Global Environmental Change*, 17(1), 47–58.

Mason, L. (1987). Tenures from Subsistence to Star Wars. In R. G. Crocombe (Ed.), *Land Tenure in the Atolls*. Fiji: Institute of Pacific Studies of the University of the South Pacific.

McAdam, J. (2010). Disappearing States, Statelessness and the Boundaries of International Law. In J. McAdam (Ed.), *Climate Change and Displacement: Multidisciplinary Perspectives* (pp. 105–129). Oxford, UK, and Portland, OR: Hart Publishing.

McArthur, P. H. (2000). Narrating to the Center of Power in the Marshall Islands. In P. Spickard & W. J. Burroughs (Eds.), *We Are a People: Narrative and Multiplicity in the Construction of Ethnic Identity* (pp. 85–97). Philadelphia: Temple University Press.

McCright, A. M. (2011). Political Orientation Moderates Americans' Beliefs and Concern about Climate Change. *Climatic Change*, 104, 243–253.

McKibben, B. (2006[1989]). *The End of Nature*. New York: Random House.

Merlin, M., Capelle, A., Keene, T., Juvik, J., & Maragos, J. (1997). *Keinikkan im Melan Aelon Kein: Plants and Environments of the Marshall Islands*. Honolulu: East-West Center, University of Hawaii.

Minnegal, M., & Dwyer, P. D. (2007). Foragers, Farmers and Fishers: Responses to Environmental Perturbation. *Journal of Political Ecology*, 14, 34–57.

Moore, E. J., & Smith, J. W. (1995). Climatic Change and Migration from Oceania: Implications for Australia, New Zealand and the United States of America. *Population and Environment*, 17(2), 105.
Mortland, C. A. (1994). Cambodian Refugees and Identity in the United States. In L. A. Camino & R. M. Krulfeld (Eds.), *Reconstructing Lives, Recapturing Meaning: Refugee Identity, Gender, and Culture Change*. Basel: Gordon and Breach Science Publishers.
Mortreux, C., & Barnett, J. (2009). Climate Change, Migration and Adaptation in Funafuti, Tuvalu. *Global Environmental Change*, 19(1), 105–112.
Moser, S. C. (2010). Communicating Climate Change: History, Challenges, Process and Future Directions. *Wiley Interdisciplinary Reviews: Climate Change*, 1, 31–53.
Moss, R. M. (2007). Environment and Development in the Republic of the Marshall Islands: Linking Climate Change with Sustainable Fisheries Development. *Natural Resources Forum*, 31, 111–118.
Murphy, S. (2007). Mantin Majol: Going, Going, Gone? *Marshall Islands Journal*, January 19, p. 13.
Namakin, B. (2008). Islands at Risk. *Marshall Islands Journal*, February 22, pp. 18–9.
Nelson, D. (2007). Expanding the Climate Change Research Agenda. *Anthropology News*, 48(9), 12–13.
Nerlich, B., Koteyko, N., & Brown, B. (2010). Theory and Language of Climate Change Communication. *Wiley Interdisciplinary Reviews: Climate Change*, 1, 97–110.
Neumann, K. (1997). Nostalgia for Rabaul. *Oceania*, 67(3), 177–193.
Neumann, K. (1992). *Not the Way It Really Was: Constructing the Tolai Past*. Honolulu: University of Hawaii Press.
Nicholls, R. J., & Cazenave, A. (2010). Sea-Level Rise and Its Impact on Coastal Zones. *Science*, 328(5985), 1517–1520.
Nicholls, R. J., et al. (2011). Sea-Level Rise and Its Possible Impacts Given a "Beyond 4°C World" in the Twenty-First Century. *Philosophical Transactions of the Royal Society A: Mathematical, Physical and Engineering Sciences*, 369(1934), 161–181.
Nickerson, R. S. (1998). Confirmation Bias: A Ubiquitous Phenomenon in Many Guises. *Review of General Psychology*, 2, 175–220.
Niedenthal, J. (2001). *For the Good of Mankind: A History of the People of Bikini and their Islands*. Majuro, Marshall Islands: Bravo Publishers.
Nilsson, C. (2008). Climate Change from an Indigenous Perspective: Key Issues and Challenges. *Indigenous Affairs*, 1–2, 8–15.
Nobel, J. (2007). Eco-Anxiety: Something Else to Worry About. *The Philadelphia Inquirer 9 April 2007.* Available at http://www.justinnobel.com/stories/ecoanxiety, philadelphia inquirer.doc
Norgaard, K. M. (2006). "We Don't Really Want to Know": Environmental Justice and Socially Organized Denial of Global Warming in Norway. *Organization & Environment*, 19(3), 347–370.
Norgaard, R. B. (2002). Optimists, Pessimists, and Science. *Bioscience*, 52(3), 287–292.
Nuttall, M. (2008). Climate Change and the Warming Politics of Autonomy in Greenland. *Indigenous Affairs*, 1–2, 44–51.
Nysta, Y. T. (2003). Manit im kwe. *Marshall Islands Journal*, May 30, pp. 10–11.
Obeyesekere, G. (1992). *The Apotheosis of Captain Cook: European Mythmaking in the Pacific*. Princeton, NJ: Princeton University Press.
OEPPC (2006). Climate Change Strategic Plan 2006. Majuro: Republic of the Marshall Islands Office of Environmental Planning and Policy Coordination (OEPPC).

Oliver-Smith, A. (2009). Climate Change and Population Displacement: Disasters and Diasporas in the Twenty-First Century. In S. A. Crate & M. Nuttall (Eds.), *Anthropology & Climate Change: From Encounters to Actions* (pp. 116–136). Walnut Creek, CA: Left Coast Press.

Oreskes, N., & Conway, E. M. (2010). *Merchants of Doubt: How a Handful of Scientists Obscured the Truth on Issues from Tobacco Smoke to Global Warming.* Bloomsbury Press.

Orlove, B. (2005). Human Adaptation to Climate Change: A Review of Three Historical Cases and Some General Perspectives. *Environmental Science and Policy*, 8(6), 589–600.

Orlove, B., Wiegandt, E., & Luckman, B. H. (2008). The Place of Glaciers in Natural And Cultural Landscapes. In B. Orlove, E. Wiegandt, & B. H. Luckman (Eds.), *Darkening Peaks: Glacial Retreat, Science and Society* (pp. 3–19). Berkeley: University of California Press.

Pacific Council of Churches (2004). The Otin'taii Declaration: The Pacific Churches Statement on Climate Change. Pacific Churches' Consultation on Climate Change, March 6–11, Tarawa, Kiribati.

Pacific Institute of Public Policy (2010). The Micronesian Exodus. Available at http://www.pacificpolicy.org/wp-content/uploads/2012/05/D16-PiPP.pdf

Parker, A., et al. (2006). *Climate Change and Pacific Rim Indigenous Nations.* Northwest Indian Applied Research Institute (NIARI). Olympia, WA: The Evergreen State College.

Parkin, D. (1999). Mementoes as Transitional Objects in Human Displacement. *Journal of Material Culture*, 4(3), 303–320.

Patel, S. S. (2006). A Sinking Feeling. *Nature*, 440.

Paton, K., & Fairbairn-Dunlop, P. (2010). Listening to Local Voices: Tuvaluans Respond to Climate Change. *Local Environment*, 15(7), 687–698.

Patt, A. G., & Schröter, D. (2007). Perceptions of Environmental Risks in Mozambique: Implications for the Success of Adaptation and Coping Strategies. Policy Research Working Paper 4417. The World Bank Development Research Group, Sustainable Rural and Urban Development Team. Available at http://www-wds.worldbank.org/servlet/WDSContentServer/WDSP/IB/2007/11/29/000158349_20071129093000/Rendered/PDF/wps4417.pdf

Penz, P. (2010). International Ethical Responsibilities to "Climate Change Refugees." In J. McAdam (Ed.), *Climate Change and Displacement: Multidisciplinary Perspectives* (pp. 151–173). Oxford, UK, and Portland, OR: Hart Publishing.

Petheram, L., Zander, K. K., Campbell, B. M., High, C., & Stacey, N. (2010). "Strange Changes": Indigenous Perspectives of Climate Change and Adaptation in NE Arnhem Land (Australia). *Global Environmental Change*, 20(4), 681–692.

Petrosian-Husa, C. C. H. (2004). Preliminary Report: Anthropological Survey of Arno Atoll. Majuro, Republic of the Marshall Islands: Historic Preservation Office.

Pettenger, M. E. (2007). Introduction: Power, Knowledge and the Social Construction of Climate Change. In M. E. Pettenger (Ed.), *The Social Construction of Climate Change: Power, Knowledge, Norms, Discourses* (pp. 1–15). Aldershot, UK: Ashgate Publishing.

Pollock, N. J. (1974). Breadfruit or Rice: Dietary Choice on a Micronesian Atoll. *Ecology of Food and Nutrition*, 3, 107–115.

Poyer, L. (1997). *Micronesian Resources Study: Marshall Islands Ethnography: Ethnography and Ethnohistory of Taroa Island, Republic of the Marshall Islands* (W. H. Adams, Ed.). San Francisco: Micronesian Endowment for Historic Preservation, Republic of the Marshall Islands, and U.S. National Park Service.

Puri, R. K. (2007). Responses to Medium-Term Stability in Climate: El Niño, Droughts and Coping Mechanisms of Foragers and Farmers in Borneo. In R.

Ellen (Ed.), *Modern Crises and Traditional Strategies: Local Ecological Knowledge in Island Southeast Asia* (pp. 46–83). Oxford: Berghahn Books.

Radio Australia (2009). Marshall Islands Continues Cleanup after Floods. Radio Australia News. Available at http://www.radioaustralianews.net.au/stories/200901/2458235.htm?desktop

Radio New Zealand (2008). British Charity Says Climate Change Breaches the Rights of Pacific Peoples. *Radio New Zealand International*, September 12. Available at http://www.rnzi.com/pages/news.php?op=read&id=41979

Radio New Zealand (2011a). Marshalls Seek UN Funding over Climate Change. *Radio New Zealand International*, January 7. Available at http://www.rnzi.com/pages/news.php?op=read&id=58052

Radio New Zealand (2011b). High Tides Flood Low-Lying Atolls in Marshall Islands. *Radio New Zealand International*, February 21. Available at http://www.rnzi.com/pages/news.php?op=read&id=58886

Radio New Zealand (2012). King Tides Force Two Marshalls Schools to Close. Radio *New Zealand International*, February 9. Available at http://www.rnzi.com/pages/news.php?op=read&id=66044

Rahmstorf, S. (2007). A Semi-Empirical Approach to Projecting Future Sea-Level Rise. *Science*, 315(5810), 368–370.

Rappaport, R. A. (1979). On Cognized Models. In R. A. Rappaport, *Ecology, Meaning and Religion* (pp. 97–144). Richmond: North Atlantic Books.

Ray, J. J. (1990). Acquiescence and Problems with Forced-Choice Scales. *Journal of Social Psychology*, 130(3), 397–399.

Rayner, S. (1989). Fiddling while the Globe Warms? *Anthropology Today*, 5(6), 1–2.

Rayner, S. (2003). Domesticating Nature: Commentary on the Anthropological Study of Weather and Climate Discourse. In S. Strauss & B. Orlove (Eds.), *Weather, Climate, Culture* (pp. 277–290). Oxford: Berg.

Rayner, S., & Malone, E. L. (1998). Social Science Insights into Climate Change. In S. Rayner & E. L. Malone (Eds.), *Human Choice and Climate Change: Vol. 4. What Have We Learned?* Columbus, OH: Battelle Press.

Reichel-Dolmatoff, G. (1976). Cosmology as Ecological Analysis: A View from the Rain Forest. *Man, New Series*, 11(3), 307–318.

Meto (2000). Republic of the Marshall Islands: Meto2000—Economic Report and Statement of Development Strategies. Manila: Asian Development Bank.

RMI Government (1996). A Situation Analysis of Children and Women in the Marshall Islands 1996. Majuro: The Government of the Marshall Islands with the assistance of UNICEF.

RMI Government (2005). Republic of the Marshall Islands Statistical Yearbook 2004. Majuro: Economic Policy, Planning and Statistics Office.

RMI Government (2006a). Coastal Management Framework—Republic of the Marshall Islands. Majuro: Republic of the Marshall Islands Environmental Protection Authority, Coastal Land Management Office.

RMI Government (2006b). Summary of Key Findings: RMI 2006: Community and Socio-economic Survey. Majuro: Republic of the Marshall Islands Economic Policy, Planning and Statistics Office.

RMI Government (2009). Views regarding the Possible Security Implications of Climate Change. Available at http://www.un.org/esa/dsd/resources/res_pdfs/ga-64/cc-inputs/Marshall_Islands_CCIS.pdf

RMI Permanent Mission (2008). National Communication Regarding the Relationship between Human Rights & the Impacts of Climate Change. UN Human Rights Council Resolution 7/23. Submitted by the Permanent Mission of the Republic of the Marshall Islands to the United Nations, December 31. Available at www.ohchr.org/Documents/Issues/ClimateChange/Submissions/Republic_of_the_Marshall_Islands.doc

Robbennolt, J. K. (2000). Outcome Severity and Judgments of "Responsibility": A Meta-Analytic Review. *Journal of Applied Social Psychology*, 30(12), 2575–2609.

Robinson, J. (2011). Radiation and Climate Change Drive Marshall Islanders to Northwest Enclave. Available at http://www.kplu.org/post/radiation-and-climate-change-drive-marshall-islanders-northwest-enclave

Rogers, R. W. (1975). A Protection Motivation Theory of Fear Appeals and Attitude Change. *Journal of Psychology*, 91, 93–114.

Roncoli, C. (2006). Ethnographic and Participatory Approaches to Research on Farmers' Responses to Climate Predictions. *Climate Research*, 33, 81–99.

Roncoli, C., Ingram, K., & Kirshen, P. (2002). Reading the Rains: Local Knowledge and Rainfall Forecasting among Farmers of Burkina Faso. *Society and Natural Resources*, 15, 411–430.

Rowa, A. (2001). Money eban korolwoj yokwe. *Marshall Islands Journal*, December 21, p. 12.

Rudiak-Gould, P. (in press). Memories and Expectations of Environmental Disaster: Some Lessons from the Marshall Islands. In M. Davies & F. Nkirote (Eds.), *Humans and the Environment: New Archaeological Perspectives for the 21st Century*. Oxford, UK: Oxford University Press.

Rudiak-Gould, P. (2009a). *The Fallen Palm: Climate Change and Culture Change in the Marshall Islands*. Saarbrücken, Germany: VDM Verlag.

Rudiak-Gould, P. (2009b). *Surviving Paradise: One Year on a Disappearing Island*. New York: Sterling Publishers.

Rudiak-Gould, P. (2010). Being Marshallese and Christian: A Case of Multiple Identities and Contradictory Beliefs. *Culture and Religion*, 11(1), 69–87.

Rudiak-Gould, P. (2011). Climate Change and Anthropology: The Importance of Reception Studies. *Anthropology Today*, 27(2), 9–12.

Rudiak-Gould, P. (2012a). Promiscuous Corroboration and Climate Change Translation: A Case Study from the Marshall Islands. *Global Environmental Change*, 22, 46–54.

Rudiak-Gould, P. (2012b). Progress, Decline, and the Public Uptake of Climate Science. *Public Understanding of Science*, published online before print June 5, 2012, doi: 10.1177/0963662512444682.

Ruel, M. (1982). Christians as Believers. In J. Davis (Ed.), *Religious Organization and Religious Experience* (pp. 9–31). London: Academic Press.

Russell, C. (2009). First Wave. *Science News*, 175(5), 24.

Sachs, J., Stege, M., & Keju, T. (2009). Recent Movements of the ITCZ in the Tropical Pacific and Ramifications for the Marshall Islands. *Earth and Environmental Science*, 6.

Sahlins, M. D. (1958). *Social Stratification in Polynesia*. Seattle: University of Washington Press.

Sahlins, M. D. (1981). *Historical Metaphors and Mythical Realities: Structure in the Early History of the Sandwich Islands Kingdom*. Ann Arbor: University of Michigan Press.

Sahlins, M. D. (1985). *Islands of History*. London: Tavistock.

Sahlins, M. D. (2005). On the Anthropology of Modernity, or Some Triumphs of Culture over Despondency Theory. In A. Hooper (Ed.), *Culture and Sustainable Development in the Pacific* (pp. 44–61). Canberra: ANU E Press.

Sarewitz, D. (2004). How Science Makes Environmental Controversies Worse. *Environmental Science & Policy*, 7, 385–403.

Schmidle, N. (2009). Wanted: A New Home for My Country. *New York Times* May 8. Available at http://www.nytimes.com/2009/05/10/magazine/10MALDIVES-t.html?_r=1&pagewanted=2&em

Sharp, M. C., Strauss, R. P., & Lorch, S. C. (1992). Communicating Medical Bad News: Parents' Experiences and Preferences. *The Journal of Pediatrics*, 121(4).

Shearer, C. (2011). *Kivalina: A Climate Change Story*. Chicago: Haymarket Books.

Silverman, M. G. (1971). *Disconcerting Issue: Meaning and Struggle in a Resettled Pacific Community*. Chicago: University of Chicago Press.

Sinclair, K. (2001). Mischief on the Margins: Gender, Primogeniture, and Cognatic Descent among the Maori. In L. Stone (Ed.), *New Directions in Anthropological Kinship* (pp. 156–174). Lanham, MD: Rowman and Littlefield.

Slovic, P., Fischhoff, B., & Lichtenstein, S. (1987). Behavioral Decision Theory Perspectives on Self-Protective Behavior. In N. D. Weinstein (Ed.), *Taking Care: Understanding and Encouraging Self-Protective Behavior* (pp. 14–42). Cambridge, UK: Cambridge University Press.

Smallacombe, S. (2008). Climate Change in the Pacific: A Matter of Survival. *Indigenous Affairs*, 1–2, 72–78.

Smith, H. A. (2007). Disrupting the Global Discourse of Climate Change: The Case of Indigenous Voices. In M. E. Pettenger (Ed.), *The Social Construction of Climate Change: Power, Knowledge, Norms, Discourses* (pp. 197–215). Aldershot, UK: Ashgate Publishing.

Smith, W. D. (2007). Presence of Mind as Working Climate Change Knowledge: A Totonac Cosmopolitics. In M. E. Pettenger (Ed.), *The Social Construction of Climate Change: Power, Knowledge, Norms, Discourses* (pp. 217–234). Aldershot, UK: Ashgate Publishing.

Southwold, M. (1983). *Buddhism in Life : The Anthropological Study of Religion and the Sinhalese Practice of Buddhism*. Manchester, UK: Manchester University Press.

Spennemann, D. H. R. (1991). *Marshallese on Eneen-kio: Evidence from Oral Traditions, Biogeography and History*. Majuro, Republic of the Marshall Islands: Alele Museum.

Spennemann, D. H. R. (1993). *Ennaanin Etto: A Collection of Essays on the Marshallese Past*. Majuro: Republic of the Marshall Islands Ministry of Internal Affairs Historic Preservation Office.

Spennemann, D. H. R. (1996). Non-traditional Settlement Patterns and Typhoon Hazard on Contemporary Majuro Atoll, Republic of the Marshall Islands. *Environmental Management*, 20(3), 337–348.

Spennemann, D. H. R. (2000). The Sea—The Marshallese World. Available at http://marshall.csu.edu.au/Marshalls/html/culture/SeaNavigation.html

Spennemann, D.H.R., & Marschner, I. (1994). Stormy Years: On the Association Between the El Niño/Southern Oscillation Phenomenon and the Occurrence of Typhoons in the Marshall Islands. Report to the Federal Emergency Management Agency Region IX, San Francisco. Available at http://marshall.csu.edu.au/Marshalls/html/typhoon/Stormy_Years.html

Sperber, D. (2000). An Objection to the Memetic Approach to Culture. In R. Aunger (Ed.), *Darwinizing Culture: The Status of Memetics as a Science* (pp. 163–173). Oxford, UK: Oxford University Press.

Sperber, D. (7969). Anthropology and Psychology: Towards an Epidemiology of Representations. In H. L. Moore & T. Sanders (Eds.), *Anthropology in Theory: Issues in Epistemology*. London: Blackwell Publishing.

Spoehr, A. (1949). *Majuro: A Village in the Marshall Islands*. Chicago: Chicago Natural History Museum.

Stege, M. (2012). When It Hits Home . . . Available at http://www.huffingtonpost.com/mark-stege/climate-change-marshall-islands_b_1443075.html

Sterman, J. D. (2011). Communicating Climate Change Risks in a Skeptical World. *Climatic Change*, 108(4), 811–826.

Stewart, P. J., & Strathern, A. J. (1997). Introduction: Millennial Markers in the Pacific. In P. J. Stewart & A. Strathern (Eds.), *Millennial Markers*.

Townsville, Australia: Centre for Pacific Studies, James Cook University of North Queensland.

Stewart, P. J., & Strathern, A. J. (1998). End Times Prophesies from Mt. Hagen, Papua New Guinea: 1995–1997. *Journal of Millennial Studies*, 1(1), N/A. Available at http://www.mille.org/publications/summer98/stewartstrathern.pdf

Stoll-Kleemann, S., O'Riordan, T., & Jaeger, C. C. (2001). The Psychology of Denial Concerning Climate Mitigation Measures: Evidence from Swiss Focus Groups. *Global Environmental Change*, 11, 107–117.

Strauss, S. (2009). Global Models, Local Risks: Responding to Climate Change in the Swiss Alps. In S. A. Crate & M. Nuttall (Eds.), *Anthropology & Climate Change: From Encounters to Actions* (pp. 166–174). Walnut Creek, CA: Left Coast Press.

Strauss, S., & Orlove, B. (2003). Up in the Air: The Anthropology of Weather and Climate. In S. Strauss & B. Orlove (Eds.), *Weather, Climate, Culture* (pp. 3–14). Oxford: Berg.

Suarez, P., & Patt, A. G. (2004). Caution, Cognition, and Credibility: The Risks of Climate Forecast Application. *Risk Decision and Policy*, 9, 75–89.

Sunstein, C. R. (2006a). The Availability Heuristic, Intuitive Cost-Benefit Analysis and Climate Change. *Climatic Change*, 77(1–2), 195–210.

Sunstein, C. R. (2006b). On the Divergent American Reactions to Terrorism and Climate Change. John M. Olin Law & Economics Working Paper No. 295, The Law School, University of Chicago. Available at http://www.columbialaw-review.org/assets/pdfs/107/2/Sunstein.pdf

Swim, J., et al. (2009). Psychology and Global Climate Change: Addressing a Multi-faceted Phenomenon and Set of Challenges. A Report by the American Psychological Association's Task Force on the Interface between Psychology and Global Climate Change. Available at http://www.apa.org/science/about/publications/climate-change-booklet.pdf

Taddei, R. (2009a). The Pragmatics of Prognostication in Times of Climate Change: Ethnographic Notes from Northeast Brazil. Paper presented at the 2009 annual meeting of the American Anthropological Association. Philadelphia, December 2, 2009.

Taddei, R. (2009b). The Politics of Uncertainty and the Fate of Forecasters: Climate, Risk, and Blame in Northeast Brazil. In V. Jankovic & C. Barboza (Eds.), *Weather, Local Knowledge and Everyday Life: Issues in Integrated Climate Studies* (pp. 287–296). Rio de Janeiro: MAST.

Tauli-Corpuz, V. (2009). Message from the UN Permanent Forum on Indigenous Issues. Ceremonial Opening of the Fifth Sami Parliament of Sweden, Kiruna, August 25. Available at http://www.un.org/esa/socdev/unpfii/documents/vtc_speech_saamiparliament09.doc

Teuatabo, N. (2002). Vulnerability of Kiribati to Climate Change and Sea-Level Rise: Scientific Information Needs to Be Comprehensive, and Unfettered of Advocacy. In J. Barnett & M. Busse (Eds.), *Proceedings of APN Workshop on Local Perspectives on Climate Change and Variability in the Pacific Islands.* December 4–6. Apia, Samoa: Macmillan Brown Centre for Pacific Studies.

Thomas, F. R. (2009). Historical Ecology in Kiribati: Linking Past with Present. *Pacific Science*, 63(4), 567–600.

Thomas, N. (1992a). Substantivization and Anthropological Discourse: The Transformation of Practices into Institutions in Neotraditional Pacific Societies. In J. G. Carrier (Ed.), *History and Tradition in Melanesian Anthropology* (pp. 64–85). Berkeley: University of California Press.

Thomas, N. (1992b). The Inversion of Tradition. *American Ethnologist*, 19(2), 213–232.

Thomas, N. (1997). *In Oceania: Visions, Artifacts, Histories.* Durham, NC: Duke University Press.

Thompson, M., Ellis, R., & Wildavsky, A. B. (1990). *Cultural Theory.* Boulder, CO: Westview Press.

Thorpe, N., Eyegetok, S., Hakongak, N., & the Kitikmeot Elders (2002). Nowadays It Is Not the Same: Inuit Qaujimajatuqangit, Climate and Caribou in the Kitikmeot Region of Nunavut, Canada. In I. Krupnik & D. Jolly (Eds.), *The Earth Is Faster Now: Indigenous Observations of Arctic Environmental Change* (pp. 198–239). Washington, DC: Arctic Research Consortium of the United States in cooperation with the Arctic Studies Center, Smithsonian Institution.

Tobin, J. E. (1952). *Land Tenure in the Marshall Islands.* Washington, DC: Pacific Science Board / National Research Council.

Tomlinson, M. (2004). Perpetual Lament: Kava-drinking, Christianity and Sensations of Historical Decline in Fiji. *Journal of the Royal Anthropological Institute*, 10, 653–673.

Tomlinson, M. (2009). *In God's Image: The Metaculture of Fijian Christianity.* Berkeley: University of California Press.

Tonkinson, R. (1977). The Exploitation of Ambiguity: A New Hebrides Case. In M. D. Lieber (Ed.), *Exiles and Migrants in Oceania.* Honolulu: University of Hawaii Press.

Toomey, C. (2009). The Maldives: Trouble in Paradise. *The Sunday Times*, February 1. Available at http://www.timesonline.co.uk/tol/news/environment/article5604464.ece

Toren, C. (1999). *Mind, Materiality, and History: Explorations in Fijian Ethnography.* London: Routledge.

Townsend, P. K. (2004). Still Fiddling while the Globe Warms? *Journal Reviews in Anthropology*, 33(4), 335–349.

Tulloch, M. (2000). The Meaning of Age Differences in the Fear of Crime: Combining Quantitative and Qualitative Approaches. *British Journal of Criminology*, 40, 451–467.

Tyler, S. A. (1986). Post-modern Ethnography: From Document of the Occult To The Occult Document. In J. Clifford & G. Marcus (Eds.), *Writing Culture: The Poetics and Politics of Ethnography* (pp. 122–140). Berkeley: University of California Press.

Uggla, Y. (2010). What Is This Thing Called "Natural"? The Nature-Culture Divide in Climate Change and Biodiversity Policy. *Journal of Political Ecology*, 17, 79–91.

UNFPA (2009). At the Frontier: Young People and Climate Change 2009. Available at http://www.unfpa.org/webdav/site/global/shared/documents/publications/2009/youth_swp_2009.pdf

Ungar, S. (2007). Public Scares: Changing the Issue Culture. In S. C. Moser & L. Dilling (Eds.), *Creating a Climate for Change: Communicating Climate Change and Facilitating Social Change* (pp. 81–88). Cambridge, UK: Cambridge University Press.

Unger, P. (1996). *Living High and Letting Die: Our Illusion of Innocence.* Oxford, UK: Oxford University Press.

United Nations (2010). World Population Policies 2009. Available at http://www.un.org/esa/population/publications/wpp2009/Publication_complete.pdf

United Nations (2011). Millennium Development Goals indicators. Available at http://mdgs.un.org/unsd/mdg/SeriesDetail.aspx?srid=749&crid

United States Census Bureau (2012). The Native Hawaiian and Other Pacific Islander Population: 2010. 2010 Census Briefs. Available at http://www.census.gov/prod/cen2010/briefs/c2010br-12.pdf

Utter, D. (1999). A Manual for Evangelism in the Marshall Islands. Thesis for doctor of ministry (D.Min.) degree. Greenville, SC: Bob Jones University.

Walsh, J. M. (1999). Jowi ko in Majol: Contemporary Significance of the Marshallese Clan System. Majuro, Republic of the Marshall Islands: Historic Preservation Office.

Walsh, J. M. (2003). Imagining the Marshalls: Chiefs, Tradition, and the State on the Fringes of U.S. Empire. Doctoral dissertation in anthropology. Honolulu: University of Hawaii.

Weaver, W., & Shannon, C. E. (1963). *The Mathematical Theory of Communication*. Chicago: University of Illinois Press.

Webb, A. P., & Kench, P. S. (2010). The Dynamic Response of Reef Islands to Sea-Level Rise: Evidence from Multi-Decadal Analysis of Island Change in the Central Pacific. *Global and Planetary Change*, 72(3), 234–246.

Weber, E. U. (2006). Experience-Based and Description-Based Perceptions of Long-Term Risk: Why Global Warming Does Not Scare Us (Yet). *Climatic Change*, 77, 103–120.

Weinstein, N. D. (1987). Cross-Hazard Consistencies: Conclusions about Self-Protective Behavior. In N D Weinstein (Ed.), *Taking Care: Understanding and Encouraging Self-Protective Behavior* (pp. 325–335). Cambridge: Cambridge University Press.

Weinstein, N. D. (1989). Effects of Personal Experience on Self-Protective Behavior. *Psychological Bulletin*, 105(1), 31–50.

Weisler, M. I. (1999). *An Archaeological Survey and Testing at Ujae Atoll, Republic of the Marshall Islands*. HPO Report 1999/05, Series Editor R. V. Williamson. Majuro Atoll, Republic of the Marshall Islands: Historic Preservation Office.

Weisler, M. I. (2001a). *On the Margins of Sustainability: Prehistoric Settlement of Utrok Atoll, Northern Marshall Islands*. Oxford, UK: Archaeopress.

Weisler, M. I. (2001b). Precarious Landscapes: Prehistoric Settlement of the Marshall Islands. *Antiquity*, 31–32.

White, G. M. (1991). *Identity through History: Living Stories in a Solomon Islands Society*. Cambridge, UK: Cambridge University Press.

Whitehouse, H. (1995). *Inside the Cult: Religious Innovation and Transmission in Papua New Guinea*. Oxford, UK: Clarendon Press.

Whitehouse, H. (2007). Towards an Integration of Ethnography, History and the Cognitive Science of Religion. In H. Whitehouse & J. Laidlaw (Eds.), *Religion, Anthropology, and Cognitive Science* (pp. 247–280). Durham, NC: Carolina Academic Press.

Whitmarsh, L. (2009). What's in a Name? Commonalities and Differences in Public Understanding of "Climate Change" and "Global Warming." *Public Understanding of Science*, 18(4), 401–420.

Whitty, J. (2003). All the Disappearing Islands. *Mother Jones*. Available at http://www.motherjones.com/news/feature/2003/07/ma_444_01.html

Wilk, R. (1995). Learning to Be Local in Belize: Global Systems of Common Difference. In D. Miller (Ed.), *Worlds Apart: Modernity through the Prism of the Local*. London: Routledge.

Wilk, R. (2009). Consuming Ourselves to Death: The Anthropology of Consumer Culture and Climate Change. In S. A. Crate & M. Nuttall (Eds.), *Anthropology & Climate Change: From Encounters to Actions* (pp. 265–276). Walnut Creek, CA: Left Coast Press.

Williamson, I., & Sabath, M. D. (1982). Island Population, Land Area, and Climate: A Case Study of the Marshall Islands. *Human Ecology*, 10(1), 71–84.

Williamson, R. V., & Stone, D. K. (2001a). Anthropological survey of Likiep Atoll. Majuro, Republic of the Marshall Islands: Historic Preservation Office.

Williamson, R. V., & Stone, D. K. (2001b). Archaeological and Anthropological Survey of Ailuk Atoll. Majuro, Republic of the Marshall Islands: Historic Preservation Office.

Wynne, B. (1992). Misunderstood Misunderstanding: Social Identities and Public Uptake of Science. *Public Understanding of Science*, 1(3), 281–304.

Xue, C. (2001). Coastal Erosion and Management of Majuro Atoll, Marshall Islands. *Journal of Coastal Research*, 17(4), 909–918.

Yamaguchi, T., et al. (2005). Excavation of Pit-Agriculture Landscape on Majuro Atoll, Marshall Islands, and Its Implications. *Global Environmental Research*, 9(1), 27–36.

Yearley, S. (2009). Sociology and Climate Change after Kyoto: What Roles for Social Science in Understanding Climate Change? *Current Sociology*, 57, 389–405.

Zedkaia, J. (2011). Keynote address. "Threatened Island Nations: Legal Implications of Rising Seas and a Changing Climate" conference, Columbia Law School Center for Climate Change Law, New York, May 23–25. Available at http://www.law.columbia.edu/null/download?&exclusive=filemgr.download&file_id=5843

Zetter, R. (2010). Protecting People Displaced by Climate Change: Some Conceptual Challenges. In J. McAdam (Ed.), *Climate Change and Displacement: Multidisciplinary Perspectives* (pp. 131–150). Oxford, UK, and Portland, OR: Hart Publishing.

Index

For Product Safety Concerns and Information please contact our EU representative GPSR@taylorandfrancis.com
Taylor & Francis Verlag GmbH, Kaufingerstraße 24, 80331 München, Germany

www.ingramcontent.com/pod-product-compliance
Ingram Content Group UK Ltd.
Pitfield, Milton Keynes, MK11 3LW, UK
UKHW052030210425
457613UK00032BA/476
* 9 7 8 1 1 3 8 9 5 2 8 1 2 *